General Preface to the Series

Because it is no longer possible for one textbook to cover the whole field of biology while remaining sufficiently up to date the Institute of Biology proposed this series so that teachers and students can learn about significant developments. The enthusiastic acceptance of 'Studies in Biology' shows that the books are providing authoritative views of biological topics.

The features of the series include the attention given to methods, the selected list of books for further reading and, wherever possible, suggestions for practical work.

Readers' comments will be welcomed by the Education Officer of the Institute.

1983

Institute of Biology
20 Queensbury Place
London SW7 2DZ

Preface to First Edition

Efficient function of cells depends to a large degree upon the constancy of their environment, therefore the composition of the extracellular fluid in all groups of animals, except the most primitive invertebrates, is the subject of complex homeostatic regulatory mechanisms: this booklet is concerned with those mechanisms. Consideration is given to control theory and the essential general principles of regulatory mechanisms as exemplified by the control of blood pressure, blood gases, hormone levels and metabolite concentrations in the mammal. In addition, more sophisticated examples of homeostasis, such as the regulation of energy intake and deep body temperature are discussed and, where appropriate, care has been taken to consider comparative aspects of physiological regulation with particular reference to the invertebrates.

I would like to express my gratitude to Carol Bradshaw and other colleagues in the University Department of Zoology for their advice on the comparative aspects of homeostasis, and to Celia Perry and Fiona Hake for their meticulous care in the preparation of the manuscript and the illustrations.

Cambridge, 1975

R. N. H.

Preface to the Second Edition

In preparing the Second Edition, the material described above has been modified where necessary to take account of recent research, but because the subject matter covered in the First Edition was of fundamental importance, nothing has been deleted. The principal innovation in this Edition has therefore been the inclusion of a substantial discussion of the homeostatic control of water and electrolyte metabolism.

I would like to express my thanks to Caroline Dawkin and Fiona Hake for their meticulous care in the preparation of the additional typescript and figures.

Cambridge, 1982   R.N.H.

Contents

Edward Arnold

© The estate of Richard N. Hardy, 1982

First published 1976
by Edward Arnold (Publishers) Limited
41 Bedford Square, London WC1B 3DQ

Reprinted 1978
Reprinted 1980
Second Edition 1983

British Library Cataloguing in Publication Data

ISBN 0-7141-2871-2

Printed and bound in Great Britain by
Spottiswoode Ballantyne Ltd
Colchester and London

1 Introduction

1.1 Historical background

Living organisms are machines: some, such as unicellular plants and animals, are relatively simple, others such as the higher mammals far exceed in complexity even the most sophisticated technological productions of man. In both living and non-living machines, however, increasing complexity must be accompanied by a parallel increase in the sensitivity of internal controlling and regulating devices: this book is concerned with that control and regulation. Obviously, it will consider principally the control of living processes, but often valuable comparisons with non-living regulating mechanisms can be made. Paradoxically perhaps, much more is known about the control mechanisms of mammals than about those of simpler animals. This reflects the fact that a great deal more research has been done in mammalian physiology, together with the fact that, by virtue of their very complexity, control mechanisms in mammals are more susceptible to detailed experimental analysis. Most of our discussion will therefore centre around the mammals, although reference will be made to regulatory systems in lower forms.

Finally, before embarking upon an examination of regulation in animal systems a brief explanatory comment on the concept of *homeostasis* is appropriate. Just over one hundred years ago the great French physiologist Claude Bernard, profoundly impressed during his research by the stability of physiological parameters such as body temperature, concluded that 'La fixité du milieu intérieur est la condition de la vie libre'. In other words, in order for an organism to function most efficiently its component cells must be surrounded by a medium of closely regulated composition.

Since his death, Bernard's principle has been overwhelmingly substantiated; it is now clear that the 'milieu intérieur' of higher vertebrates is the subject of a multiplicity of complex regulatory mechanisms and, in consequence, its composition is controlled to within very fine limits. In 1929, Walter B. Cannon, the American physiologist, coined the term 'Homeostasis' to describe this state of affairs: its original application is perhaps best explained in his own words: 'The constant conditions which are maintained in the body might be termed *equilibria*. That word, however, has come to have fairly exact meaning as applied to relatively simple physicochemical states, enclosed systems where known forces are balanced. The co-ordinated physiological processes which maintain most of the steady states in the organism are so complex and so peculiar to living beings – involving as they may, the brain and nerves, the heart, lungs, kidneys and spleen, all working co-operatively – that I have suggested a special designation for these states, *homeostasis*. The word does not imply something set and immobile, a stagnation. It means a condition – a condition which may vary, but which is relatively constant.' The word 'homeostasis' is derived from two Greek words '*homeo*' = similar and '*stasis*' = standing: in this book it will be used both as a description of the regulated state and as indicative of the regulatory mechanisms implicit in such a state.

1.2   Essentials of regulatory systems

The statement that particular physiological variables, such as deep body temperature and the pressure, pH and glucose concentration of blood, are subject to homeostatic regulation carries with it certain implications. First, means must exist to monitor changes in the variable regulated (input). Secondly, there must be some means of interpreting and integrating this sensory information in order to produce appropriate corrective responses, and finally, effector mechanisms (output) must be present to counteract externally induced changes in the variable sufficient to remove it beyond the acceptable limits.

Figure 1–1 summarizes in the case of higher vertebrates the interrelationships between these three basic components of control systems. Integration devolves almost exclusively upon the central nervous system (CNS). Sense organs outside the CNS (called for convenience 'peripheral' sense organs) transmit information about changes in both external and internal environment to the CNS via afferent (sensory) nerve

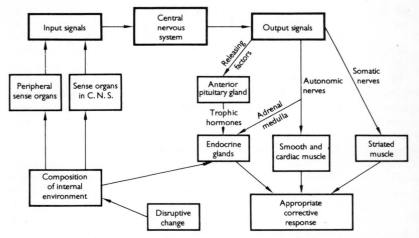

Fig. 1–1   Diagram to illustrate the interrelations between input and output signals and the central nervous system.

fibres. The CNS itself also contains many regions sensitive to specific changes in blood composition ('central receptors'). Information from both peripheral and central sense organs composes the 'input signal' to the CNS. Acting on this information, the CNS produces an appropriate output signal which induces corrective responses from the effector tissues. Output signals, which may be carried by both nerves and hormones, comprise the subject matter of Chapter 2.

1.3   Control theory

Homeostasis is the term used to describe regulation of the internal environment of living organisms: the analogous term in the case of machines is servomechanisms. Wiener in 1948, realizing that the two types of control systems had many aspects in common, coined the term *cybernetics* (Greek = steersman) to describe the general study of control mechanisms: this is often also called 'control theory'.

Control theory is essentially directed towards defining precisely, usually in mathematical terms, the behaviour of a complex control system. In theory, although rarely in practice in the case of biological systems, this object can be achieved if one can delineate accurately the properties of the component parts of the control system. However, even in the case of an extremely simple physical system, precision in predicting the behaviour of the system is difficult to achieve. Consider the example of a simple electrical circuit containing a battery in series with a variable resistance: Ohm's Law would seem to be the only consideration in a cybernetic analysis of factors relating variation in resistance to current flow; thus current flow could be calculated by dividing the battery EMF by the resistance of the circuit. However, closer analysis would reveal at least two time-dependent factors that would complicate analysis of the behaviour of the circuit: the EMF of the battery would gradually decrease while the resistance of the circuit would tend to increase as the temperature of the resistance elements increased. Clearly, careful measurements on comparable circuits could result in the production of mathematical corrections for these latter factors and, eventually, one could produce a complex analysis of the behaviour of the system under closely defined conditions. It should be apparent from this simple example that a *precise* cybernetic analysis of even a crude two-component system is extremely complex – even when the laws governing behaviour of the system are well established. Consider how much more involved the analysis of a homeostatic mechanism must be, particularly when it is remembered that the laws governing biological processes are largely unknown, usually non-linear and certainly somewhat more complex than Ohm's Law! Much current research is now in fact directed towards the production of computer-simulations of physiological control systems. Computers are programmed with data obtained from experimental observations and are then asked to derive the mathematical relation between stimulation (input) and response (output) under one set of experimental conditions. Further experiments will then determine whether the calculated relationship holds under different conditions. If, as is usually the case, it doesn't, then the computer modifies its equations until, by a process of trial and error, one obtains a mathematical model which simulates with reasonable accuracy the behaviour of the homeostatic system. Even today such models are usually both imperfect and incredibly cumbersome; nevertheless, the process of developing them often leads to fruitful future lines of practical experimental analysis.

1.4 Feedback

Perhaps the most important concept to have arisen from control theory studies is that of *feedback*. Feedback is difficult to define without recourse to diagrams and examples, so we will consider specific examples of both *negative* and *positive* feedback and then derive a general definition.

Consider the example of a simple, thermostatically-controlled laboratory water-bath with the thermostat set to 37°C. The water in the bath will lose heat to its surroundings and thus cool down but, as soon as the temperature drops below 37°C, the heater is activated and raises the temperature to 37°C again – at which point the heater is turned off and the cycle restarts. This is a simple example of negative feedback: the consequence of activity of the heater, i.e. the increase in the water temperature, is itself responsible for turning the heater off. Expressed in cybernetic terms the output of the system determines the input. However, it does so in a precisely determined way: namely by reducing the error between the actual bath temperature (<37°C) and the desired 'set-point' temperature (37°C); when the error is zero the output ceases. Perfect control is, in essence, the abolition of error: practical control systems are designed to minimize error. In the example above, the error is a water temperature different from the 'set-point' and the negative feedback control of the heater, through its thermostat, serves to correct it if the temperature falls *below* 37°C. You will note, however, that it can do nothing to prevent the water temperature exceeding 37°C – apart of course from the fact that current flow through the heater is terminated. In practice, in simple on/off thermostatic controls, residual heat from the hot elements raises the temperature transiently above 37°C even after current flow has ceased. Similarly, there is a brief time lag between when the heater is activated and the time when the water temperature begins to increase. The oscillation in water temperature on either side of 37°C is an example of 'hunting' in a control system.

More sophisticated control systems, and in fact most biological systems, do not work with a simple on/off or 'all-or-nothing' output; there is an element of proportional control. Thus, as the error decreases, so does the output, which gradually 'tapers off' to zero when the error is zero. In such systems 'hunting' is minimized, but the nature of the 'black box' which relates input to output is much more complex – expensive in the case of machines and, to the chagrin of investigating scientists, much more perplexing in biological contexts!

Negative feedback in some form or other can be found incorporated into most, if not all, homeostatic mechanisms; examples discussed in later chapters include the reflex regulation of arterial blood pressure, regulation of gas tensions and hormone levels in blood, and the stabilization of deep body temperature. Each of these parameters is precisely maintained at an optimal value (often called the 'set-point') analogous with our 37°C water bath: deviation from this value sets in train complex regulatory mechanisms which exemplify the principles of negative feedback.

While negative feedback is associated with regulation or stabilization of physiological variables, positive feedback has precisely the reverse effect. Here the portion of the output fed back to the controller tends to exacerbate rather than to reduce the error: the process thus enters an irreversible phase – a sort of 'vicious circle'. By their very nature, examples of positive feedback are less easy to come by, both in inanimate systems and in living organisms. Consider an unskilled driver occupying a loose seat in a rapidly moving car. If, in these unlikely circumstances, the driver gently applied the brake, the deceleration would propel seat and occupant forward, thereby applying the brake harder and increasing the decelerating force until eventually the car would come to an exceedingly abrupt halt! Simple combustion could be construed as positive feedback; increasing temperature facilitates further combustion, a process analagous with a chain reaction in nuclear physics. A biological example of positive feedback occurs at the threshold stage of development of a nerve impulse; the initial depolarization increases the entry of Na^+ into the negatively charged nerve interior, thereby causing further depolarization and thus further Na^+ entry . . . , until at the height of the nerve spike the process ceases because of a delayed cut-off of the Na^+ access channels, together with the development of an opposing electrochemical force due to positive charges within the fibre.

Feedback then is a phenomenon in which the consequences of the action of a system (output) are referred back to the system as part of its input, thereby influencing its subsequent output. Such a system can therefore appreciate what it has achieved, compare it with what it should have achieved and, in the case of negative feedback, modulate its output appropriately to minimize the error between target and achievement.

The foregoing account is a necessarily brief and deliberately non-mathematical introduction to the most elementary principles of control theory. It is a fascinating subject and would fully justify examination of the further reading suggested on p. 66.

2   Effector Mechanisms

2.1   Nervous pathways

As outlined in Chapter 1, the nerves which carry executive information from the central nervous system to the effector organs of mammals are of two types; axons supplying the contractile elements of skeletal muscle (somatic nerves) and axons supplying all other effector tissues (autonomic nerves).

The somatic system demonstrates extreme sophistication, both in the interrelationship between command signals from many parts of the brain and in their modification by proprioceptive information from muscles and joints. In consequence, voluntary movements are smooth and precise and appropriately co-ordinated with the requirements of balance and posture. However, such mechanisms are not strictly relevant to a discussion of homeostasis as defined in Chapter 1 and will not be considered further.

2.1.1   The autonomic nervous system of mammals

The autonomic nervous system comprises the innervation which controls smooth muscle, cardiac muscle, exocrine glands (glands which release their secretion into ducts, e.g. salivary and sweat glands) and certain endocrine glands. (N.B. secretion from most endocrine glands is not controlled by nerves.) It is obvious therefore that the autonomic nervous system must be intimately involved in almost all aspects of homeostasis by virtue of its profound influence on the effector tissues of the body.

The autonomic nervous system comprises two divisions: sympathetic and parasympathetic. This separation was made originally on anatomical criteria but can also be supported on both functional and pharmacological gounds. Figure 2–1 represents the general arrangement of the two divisions of the autonomic nervous system in man. The parasympathetic system has a relatively restricted distribution; its fibres are found in certain cranial nerves and in the spinal nerves from the sacral segments (hence the alternative name 'cranio-sacral' division). Parasympathetic nerves are only found in organs within the head and trunk: there are none in the appendages. Indeed, with the exception of those in the vagus which extend from the brain down into the thorax and abdomen (vagus = Latin for wanderer), most parasympathetic fibres are short and circumscribed. In contrast, sympathetic fibres are found in almost every part of the body. They originate in the thoracic and upper lumbar segments of the spinal cord ('thoracico-lumbar' division) and, in addition to supplying those specific organs shown in Figure 2–1, ramify

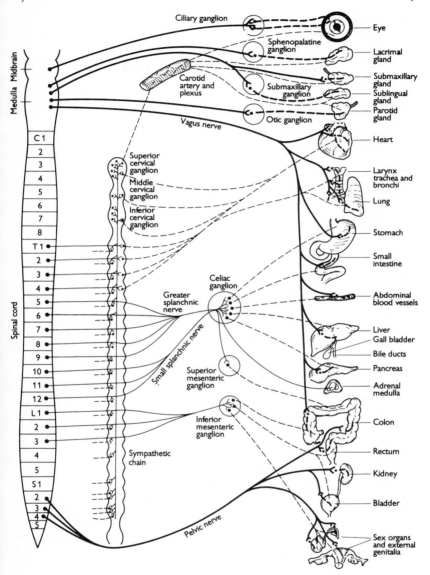

Fig. 2–1 The autonomic pathways in man. Preganglionic neurones shown as continuous lines; postganglionic neurones as dotted lines. Heavy lines are parasympathetic fibres; light lines show sympathetic fibres. (From YOUMANS, W. B. (1962), *Fundamentals of Human Physiology*, Year Book Medical Publishers.)

throughout the body in association with virtually every blood vessel except true capillaries. Furthermore, their presence in most parts of the skin, in conjunction with sweat glands and the piloerector muscles attached to hair follicles, is vital to thermoregulation (see Chapter 5).

It is often thought that the sympathetic and parasympathetic systems are mutually antagonistic, but, while this is certainly true in some instances (sympathetic nerves increase heart rate, vagal fibres slow the heart), there are many circumstances where the two systems may act to produce a common end (the control of salivary and pancreatic secretion). Moreover, as mentioned above, many tissues have no parasympathetic supply: nevertheless, their activity can be both augmented and diminished by their sympathetic nerves alone. How is this achieved? The answer lies in the resting or 'tonic' activity present in most sympathetic nerves: this principle is demonstrated in Fig. 2–2 by reference to those sympathetic nerves which cause contraction of the smooth muscle of

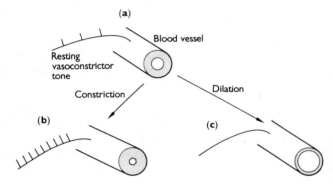

Fig. 2–2 Tonic activity in sympathetic vasoconstrictor nerves. (a) Resting vasoconstrictor tone (1–5 nerve impulses/s) holds blood vessel partially constricted. (b) Increase in nerve discharge frequency causes further constriction. (c) Abolition of nerve activity allows dilation.

blood vessels (vasoconstrictor fibres). Under resting conditions, a continual stream of nerve impulses arrives at the blood vessel causing it to remain in a partially constricted state: this is called vasomotor tone and plays an essential part in the maintenance of normal blood pressure (see Chapter 3). The vessel can be caused to constrict further by increasing the frequency of the sympathetic nerve impulses, but on the other hand it can be made to dilate by decreasing the resting tonic discharge of impulses. Tonic activity is widespread in the sympathetic system but rare in the parasympathetic: it originates within the central nervous system, particularly in certain integrating centres within the medulla and hypothalamus (see e.g. pp. 23, 25, 51).

A second common misconception about autonomic activity is that the sympathetic system is principally an emergency device functioning under circumstances of 'fight or flight'. While it is certainly true that it plays a vital role at such times by activating those cardiovascular and metabolic changes which compose the 'sympathetic discharge' emergency response, it is important to remember that it is at least as important to normal function under more placid conditions.

2.1.2 Chemical transmission

Before leaving the subject of the mammalian autonomic nervous system, mention must be made of the chemical transmission of information at the nerve endings, since an understanding of this will aid subsequent interpretation of autonomic components of homeostasis. It will be seen from Figs 2–1 and 2–3 that, unlike the single motor neurones of the somatic system, the autonomic motor pathway comprises two neurones. The first (preganglionic) fibres are myelinated, have their cell bodies in the CNS and terminate in ganglia composed of the cell bodies of the second (postganglionic) neurones whose unmyelinated fibres then run to the effector organ. Parasympathetic preganglionic fibres are relatively long and liberate acetylcholine (ACh) at ganglia which are usually found close to or within the effector organs. In consequence, the postganglionic parasympathetic fibres are extremely short: they too release ACh. Sympathetic preganglionic fibres are usually short and most terminate in a well-defined string of lateral ganglia close to the spinal cord (sympathetic chain) where they release ACh. Sympathetic postganglionic fibres are relatively long and it is they that ramify extensively to provide the system with its ubiquitous distribution. Unlike all other autonomic neurones, they release noradrenaline at their endings (certain anomalous sympathetic postganglionic fibres which dilate muscle blood vessels and control thermal sweating release ACh).

The adrenal medulla can be considered as embryologically equivalent to a sympathetic ganglion minus axons and, as such, is supplied by normal sympathetic preganglionic nerves and in response to them will release noradrenaline. However, in most adult mammals it also releases adrenaline in greater quantities since, unlike normal postganglionic cells, it possesses an enzyme which will methylate noradrenaline and thereby produce adrenaline. Both adrenaline and noradrenaline are released from the adrenal medulla directly into the blood, making the adrenal medulla by definition an endocrine gland. It is of historical interest that similarities between injections of adrenal medulla extract and sympathetic nerve activity led in 1904 to the first suggestion of chemical transmission at nerve endings. The 'mass-production' of sympathetic transmitter by the adrenal medulla greatly aids the concerted discharge of the system in emergency. In contrast, the discrete function of components of the parasympathetic system would be severely prejudiced if there were mass release of ACh, not to mention the action of striated muscle where neuromuscular transmission also involves ACh. In fact, once it is released from the nerves and has performed its function, ACh is rapidly destroyed by cholinesterases concentrated at the nerve endings, and any that escapes is broken down by similar enzymes in the blood. Indeed many of the so-called 'nerve gases' developed during and after the Second World War, depend for their effect upon blocking the action of the cholinesterases. The debilitating and ultimately fatal effect of this can be envisaged by considering the consequences if *all* the nerves releasing ACh fired maximally and continuously.

Finally, there is growing evidence that other chemicals, in addition to ACh and noradrenaline, may mediate transmission at certain autonomic synapses. This branch of the science of neuropharmacology is still in its infancy, but there is already an impressive list of possible 'new' transmitters which includes VIP (vasoactive intestinal peptide released by parasympathetic nerves in certain salivary glands) dopamine (a widely distributed precursor of noradrenaline itself) and a number of amino acids such as glycine and glutamine which probably are restricted to actions within the CNS.

2.1.3   Agents that block chemical transmission

Synapses at which the electrical energy of the presynaptic nerve impulse promotes the release of a chemical transmitter may be blocked by drugs which interfere with release of the transmitter or its reaction with the postsynaptic cell. The properties of the receptor sites on the postsynaptic membrane differ widely so that it is possible to block, for example, ACh mediated transmission at postganglionic parasympathetic endings without affecting either autonomic presynaptic transmission or transmission at somatic motor end-plates. Such differential blockade is of extreme value in medicine and has considerable application in the physiological analysis of the component parts of many complex homeostatic mechanisms. (Some important common blocking drugs and their sites of action are included in Fig. 2–3.)

2.1.4   Visceral nerves in submammalian species

Most submammalian species possess nerves which supply the viscera, as distinct from those nerves supplying the locomotor organs. A visceral plexus is present in many annelids, arthropods and molluscs and there is a definite innervation of the pharynx and rectum in nematodes, and of the pharynx and copulatory organs in flatworms. A common arrangement is a circumenteric nerve ring at the anterior end, and sometimes the posterior end, of the gut with branching nerves from it. In invertebrates, the supply is primarily motor to the visceral musculature and this seems to have been its original role – glands, chromatophores etc., coming under autonomic control at a later stage. Certainly insects and crustacea appear to lack nervous control of secretion into the gut, but the

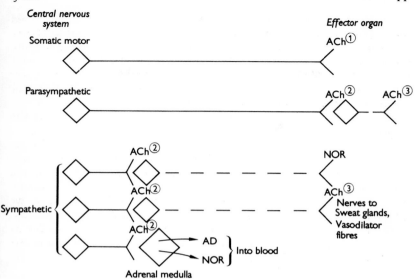

Fig. 2–3 Chemical transmission in the mammalian peripheral nervous system. ACh, acetylcholine; NOR, noradrenaline; AD, adrenaline. ① blocked by curare, ② blocked by ganglion-blocking drugs (e.g. hexamethonium), ③ blocked by atropine.

earthworm is noteworthy in that stimulation of visceral nerves will result in the secretion of proteolytic enzymes into the gut.

Details of the control mechanisms are less well understood in the lower groups than in more advanced invertebrates where the system usually depends on a balance of excitation and inhibition by different transmitters, but generalisations are of little value as the details vary considerably even between species. Thus, most molluscan hearts are slowed by acetylcholine but the rate of beat in one species of mussel is increased. Acetylcholine, adrenaline and 5-hydroxytryptamine are found in the invertebrates and can all affect visceral functions but the evidence for their doing so under physiological conditions is much less good than that for vertebrates.

The anatomical arrangement of the system in *Amphioxus* is similar to that of many pre-vertebrate groups, consisting of motor fibres from the central nervous system to the nerve plexus of the viscera. Little is known of the details of control or the transmitters used. The fibres in the central nervous system can be thought of as corresponding to the preganglionic, and the cells of the nerve net to the postganglionic fibres in vertebrates; but there is no division into parasympathetic and sympathetic and no innervation of the skin.

In the lower vertebrates, various parts of the typical fully developed mammalian-type autonomic nervous system may be present. In *Myxine* (hag-fish) the parasympathetic is represented only by the vagus and there is no cardiac innervation; there seem to be some segmental visceral fibres representing the sympathetic but no ganglia have been definitely identified. *Lampetra* (lamprey) has a vagal supply to the heart and a rather better developed sympathetic system with diffuse ganglia.

In fish, there is no sacral division of the parasympathetic and the cranial division is often incomplete. Elasmobranchs lack a sympathetic supply to the skin or head. Furthermore, the vagus does not affect the gut and there appears to be no autonomic control of glands. By contrast, teleosts have sympathetic innervation to the skin and head and a parasympathetic supply to the gut, but on the whole the autonomic system of fish is less clearly separated either physiologically or pharmacologically, into the two divisions seen in the mammal.

In amphibia, reptiles and birds the mammalian-type pattern is established; the parasympathetic system has a sacral division and the oesophagus, stomach, rectum, bladder and heart all have the typical double innervation of the mammalian system.

2.2. Endocrine pathways

There is a second route, in addition to nerves, by which executive information may be transmitted from one part of an organism to another: it involves the production of specific chemical 'messengers' called hormones (from the Greek 'I excite'). Hormones may conveniently be defined as substances secreted in response to a specific stimulus by circumscribed groups of cells (endocrine organs) and carried to the target organ or organs in the blood or other circulating body-fluid. Hormones are not used as energy sources by the target organs but serve to regulate reactions in the organ or tissue. The essential difference between transmission of information by nerves and that by humoral (hormonal) pathways is that with the former there are specific *cellular* links between the communicating tissues: in the latter case, the communicating link is the circulating body-fluid and consequently the hormone in fact reaches all parts of the body, although its action may be restricted to a particular target gland by a specific chemical affinity for its cells. Table 1 lists the principal mammalian hormones, their source(s) and their sites of action.

It is a reasonable generalization that nervous pathways are primarily of importance in controlling processes in which a rapid effector response is essential or in which continuous short term modulation of response is desirable. Endocrines are better suited for more protracted regulatory tasks in which sudden alterations in response are unnecessary. However, notwithstanding the above generalization, it is becoming increasingly apparent that it is naïve to attempt a rigid separation of nervous and endocrine regulation. The study of neuroendocrinology is expanding rapidly.

Table 1   The principal mammalian hormones.

Source	Hormone	Site of action
Hypothalamus	Corticotrophin releasing factor	Anterior pituitary (ACTH)
	Thyrotrophin releasing hormone	Anterior pituitary (TSH)
	Growth hormone releasing factor	Anterior pituitary (GH)
	Growth hormone inhibitory hormone (somatostatin)	Anterior pituitary (FSH)
	Gonadotrophin releasing hormone (GnRH)	Anterior pituitary (LH)
	Prolactin inhibitory factor	Anterior pituitary (Prolactin)
Hypothalamus/ posterior pituitary	Antidiuretic hormone (Vasopressin)	Kidney
	Oxytocin	Uterus and mammary gland
Anterior pituitary	Adrenocorticotrophic hormone (ACTH)	Adrenal cortex
	Thyroid stimulating hormone (TSH)	Thyroid
	Growth hormone (GH)	Most tissues
	Follicle stimulating hormone (FSH)	Ovary or testis
	Luteinizing hormone (LH)	Ovary or testis
	Prolactin	Ovary and mammary gland
Thyroid	Thyroxine	Most tissues
	Calcitonin	Most tissues
	Thyrocalcitonin	Bone and kidney
Parathyroid	Parathormone	Bone and kidney
Pancreas	Insulin	Most tissues
	Glucagon	Most tissues
Adrenal cortex	Glucocorticoids (cortisol and corticosterone)	Most tissues
	Mineralocorticoids (aldosterone)	Kidney and exocrine glands
Adrenal medulla	Adrenaline	Most tissues
	Noradrenaline	Most tissues
Ovary/placenta	Oestrogens	Uterus, mammary gland, brain
	Progesterone	Uterus, mammary gland, brain
Gastrointestinal tract	Gastrin	Stomach
	Secretin	Pancreas
	Pancreozymin/cholecystokinin	Pancreas
Kidney	Renin	Blood
	Erythropoietin	Bone marrow

2.2.1 Regulation of endocrine activity

We have seen that there are two principal executive communication systems in the body, nerves and hormones, so it is important to explore how far these two systems interact and in particular to appreciate the degree to which endocrine activity is under the control of the central nervous system in mammals.

Certain endocrines appear to function with no reference at all to the nervous system. The regulation of blood calcium concentration depends on a simple direct effect of blood calcium level on the two hormones concerned. Fall in blood calcium stimulates the release of the parathyroid hormone (parathormone) which then mobilizes calcium stores and increases the blood concentration. Excess blood calcium stimulates the release of calcitonin from the thyroid gland which has effects antagonistic to parathyroid hormone and therefore lowers blood calcium. The interaction of these two hormones thus ensures that the free calcium level in blood is maintained within closely circumscribed limits (Fig. 2–4a).

The endocrine pancreas (islets of Langerhans) provides another example of a gland little influenced by the nervous system. The principal controlling factor for pancreatic endocrine secretion is the blood glucose concentration (Chapter 4): increase stimulates insulin secretion and decrease stimulates glucagon release. Each hormone tends to produce a change in blood glucose concentration in the direction which would restore it to within the required range. However, in addition to the direct effect of blood glucose on the pancreatic islets, there is now evidence to indicate that both the sympathetic and the parasympathetic nerves can also influence insulin and glucagon secretion.

The adrenal medulla provides an extreme contrast to the parathyroid and endocrine pancreas (Fig. 2–4b), because its activity is almost entirely dependent upon direct nervous control from the sympathetic preganglionic fibres supplying it (p. 9). There are two types of cell in the medulla secreting adrenaline and noradrenaline respectively and it seems probable that the two hormones have relatively independent nervous control. Low blood oxygen or low blood glucose can also stimulate the gland directly, although this is probably of little significance except in the foetus and new-born animal when the innervation may not be fully functional.

The posterior pituitary gland and the adrenal medulla are the only endocrine glands *directly* controlled by nerves. However, there is an important difference in the way in which nerves affect the glands. The adrenal medulla has a simple *secretomotor* innervation entirely comparable with that controlling the activity of exocrine glands. The posterior pituitary is itself a downgrowth of the brain and appears simply to serve as a point where secretions produced in certain parts of the hypothalamus are released into the blood. In this case therefore the nerve axons

(a) *Direct control by a constituent in blood*

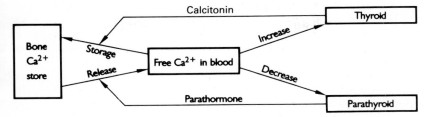

(b) *Direct control by nerves*

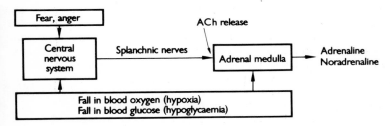

(c) *Neurosecretion*

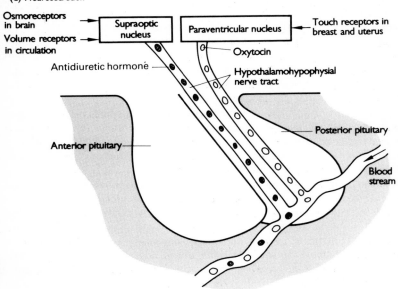

Fig. 2–4   Examples of simple endocrine control mechanisms (see text).

connecting the hypothalamus to the posterior pituitary act simply as pipes along which the hormones pass. Antidiuretic hormone (vasopressin) and oxytocin, the so-called posterior pituitary hormones are thus really hypothalamic hormones synthesized by the cell bodies of the neurones composing the nerve tracts connecting hypothalamus to posterior pituitary. The hormones pass down the nerve fibres in association with carrier proteins called neurophysins by a process of 'axonal flow' and are released into capillaries within the posterior pituitary: the whole mechanism is known as 'neurosecretion' (Fig. 2–4c).

So far, the glands under nervous control we have discussed have had an obvious connection with the central nervous system—nerves! There are, however, many glands, such as the thyroid, adrenal cortex and gonads whose activities are related to external environmental changes such as cold, stress, day length, etc. These glands must clearly be influenced by the nervous system, but, since they function equally well when transplanted, they cannot depend on direct nervous connections. Such glands are now known to be controlled by trophic hormones from the anterior pituitary (Table 1) so the question of nervous control devolves upon how the nervous system controls the anterior pituitary. Unlike the posterior pituitary, the anterior pituitary is not derived from nervous tissue but is in fact an outgrowth of the buccal cavity. Nevertheless, it was thought for many years that there were secretomotor nerves running from hypothalamus to anterior pituitary. This idea has now been disproved and it has been established instead that the functional connection is a vascular one; the hypothalamo-hypophyseal portal system. The blood flowing in this system carries specific chemical factors from the primary capillary bed in the hypothalamus down to the secondary capillary bed in the anterior pituitary. These factors (collectively known as *hypophysiotrophic factors*) promote or inhibit the release of one of the six anterior pituitary hormones and are named in a way that describes their action (also, by convention, those which have been identified and chemically characterized are accorded the status of 'hormones' – the rest remain 'factors'); e.g. thyrotrophin releasing hormone (TRH), prolactin inhibiting factor (PIF). The six most important factors are shown in Table 1. They are present in extremely small amounts in the hypothalamus and portal blood and are virtually undetectable in peripheral blood. Nevertheless, three of them have been identified chemically; TRH (a tripeptide), growth hormone inhibitory hormone also called somatostatin (a peptide with 14 amino acids) and gonadotrophin releasing hormone (a peptide with 10 amino acids). Knowing the structure of these simple peptides allows the synthesis of chemical analogues which may have important medical applications. For example, synthetic gonadotrophin releasing hormone and its analogues are used in the treatment of certain forms of infertility. Moreover, somatostatin, which has been described as 'endocrine cyanide', blocks the secretion of

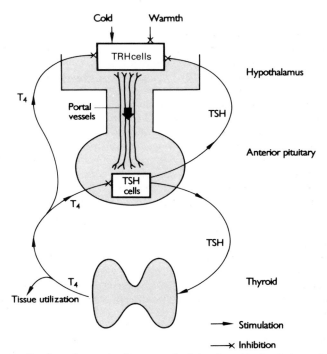

Fig. 2–5   Feedback pathways in the control of the thyroid gland T_4, thyroxine. TSH, thyroid stimulating hormone. TRH, thyrotrophin releasing hormone.

other hormones in addition to growth hormone. Therefore it, and some of its analogues hold promise therapeutically as a means of controlling abnormal secretion of the endocrine pancreas and certain gastro-intestinal hormones. Not all hypophysiotrophic factors may transpire to be peptides: there is growing evidence that the prolactin inhibitory factor is dopamine.

2.2.2   Hypothalamus–anterior pituitary–target gland interactions

The interactions between the hypothalamus, the anterior pituitary and a 'target gland' such as the thyroid are complex and provide interesting examples of homeostatic principles. At the first level of analysis the system is relatively simple (Fig. 2–5): thyroid stimulating hormone (TSH) from the anterior pituitary stimulates thyroid secretion of thyroxine (T_4), T_4 in the circulation then inhibits TSH secretion. This is a classical negative feed-back loop and would serve to stabilize blood T_4 concentration at a pre-set level.

The T_4 in the blood is continually being utilized at the tissue level so that there is a steady secretion of T_4 from the thyroid. Increased tissue utilization is reflected in a decrease in T_4 inhibition of the pituitary and therefore an increased TSH secretion sufficient to stimulate the thyroid to restore blood T_4 concentration. However, further examination of the system reveals additional feedback pathways: T_4 also inhibits the hypothalamic secretion of TRH and thus indirectly reduces TSH secretion, while TSH secretion from the pituitary may be itself subject to negative feedback control since it may act on the hypothalamus to inhibit TRH secretion. Finally, superimposed upon all these feedbacks is the neural control of TRH secretion whereby factors such as environmental temperature can influence TRH secretion directly and thereby 'reset' the thresholds for the feedback mechanisms.

The control of T_4 secretion is thus geared both to the maintenance of steady state conditions and to appropriate modulation of the steady state depending on the animal's requirements. To revert to a waterbath analogy, T_4 feedback at the hypothalamic and pituitary level and TSH feedback at the hypothalamus are comparable to the basic thermostat ensuring sensitive control of a preset waterbath temperature, while the direct neural control of TRH adjusts the setting of the thermostat.

Before leaving our discussion of endocrine regulation it is worthwhile mentioning an example of *positive* feedback, although of course this does not fall strictly within the compass of this book. The anterior pituitary secretes luteinizing hormone (LH), which stimulates both the secretion of oestrogen from the ovarian follicles and also, in high concentrations, will cause ovulation. Oestrogen meanwhile will stimulate the secretion of the hypothalamic releasing hormone for LH (GnRH). Just prior to ovulation GnRH secretion begins to increase due to the action of the neural mechanisms determining the time of ovulation. In turn, the increase in GnRH stimulates LH which then promotes oestrogen secretion from the ovary. The oestrogen rapidly amplifies the initial GnRH increase by its stimulant action on the hypothalamus. Thus, more LH is released, stimulating more oestrogen in a self-regenerative cycle. Eventually LH concentrations become sufficient to cause ovulation. The cycle is then broken by the intervention of progesterone secreted by the corpus luteum formed in the follicle after ovulation. Progesterone feeds back on the hypothalamus to inhibit the release of GnRH and thus decreases LH secretion. It is worth noting that the feedback inhibition of LH by progesterone is of great practical importance, since the high levels of ovarian and placental progesterone during pregnancy, by inhibiting LH secretion, serve to ensure that further ovulation does not occur.

2.2.3   *Endocrine regulation in sub-mammalian species*

Amongst vertebrates there is general conformity in the way in which the principal endocrine glands are organized. Thus birds, reptiles, amphibia

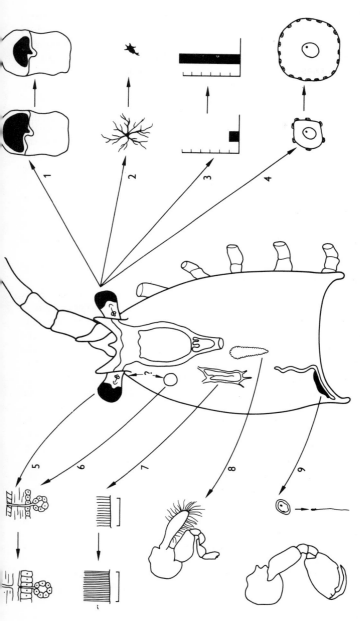

Fig. 2–6 Diagram to summarize the principal hormonal functions in crustacea. On the right side 1–4 represent hormones from the eyestalk; on the left, 5–9 represent eyestalk and other endocrine effects. 1, light-adapting distal retinal pigment hormone; 2, chromatophorotropins; 3, hyperglycaemic hormone; 4, eyestalk ablation results in ovarian growth, through precocious vitellogenesis; 5, moult-inhibiting hormone of the eyestalk probably acting normally on the Y-gland; 6, the Y-gland from which is secreted a moult hormone; 7, the pericardial organ, extracts of which accelerate heart rate; 8, ovarian hormones regulating female secondary sex characters; 9, androgenic gland of the male, regulating spermatogenesis and secondary sex characters in the male. (From CHARNIAUX-COTTON, H. and KLEINHOLZ, L. H. (1964). *The Hormones*, Vol. IV, p. 135, Academic Press.)

and fish all possess thyroid and adrenal glands and endocrine tissue is present in the pancreas and gonads, although the hormones secreted by these glands and their precise functions may differ between groups. For example, thyroid hormones are important in amphibian metamorphosis and fish osmoregulation and adrenal glucocorticoids have an osmoregulatory function in fish which do not secrete the mineralocorticoid aldosterone. Moreover, fish do not possess parathyroid glands and in submammalian vertebrates calcitonin is secreted by the ultimobranchial bodies; endocrine glands now incorporated in the mammalian thyroid. Teleost fish also have a neurosecretory endocrine organ in the caudal region of the spinal cord, the urophysis. This gland secretes substances which affect sodium and water movements, bladder contraction and blood pressure: it is thought to play a role in osmoregulation. The anterior and posterior pituitary are both present in submammalian vertebrates and provide, as in the mammal, an important link between neural and endocrine effector pathways. In addition, the intermediate lobe (pars intermedia), an area of uncertain function in mammals and birds, provides an important hormone, melanocyte-stimulating hormone (MSH), which influences pigment dispersion in the skin of certain fish, amphibia and reptiles.

The endocrinology of a few invertebrate groups, notably the annelids, crustacea and insects has been extensively studied and much is known of the intricate control of metamorphosis and reproduction. However, it is inappropriate to consider invertebrate endocrinology in detail here since only rarely have invertebrate hormones been shown to manifest a truly homeostatic function. Nevertheless, it is instructive to appreciate the extent to which endocrine mechanisms have been defined in certain invertebrate groups: Fig. 2–6 summarizes the principal hormonal mechanisms which have been established in the crustacea.

One final point before leaving our discussion of effector pathways concerns the very widespread distribution of certain neural transmitter substances within the animal kingdom. Acetylcholine, adrenaline and 5-hydroxytryptamine have well-established transmitter actions within the peripheral and central nervous systems of vertebrates. Acetylcholine also occurs in some protozoa and in almost all invertebrate groups. Catecholamines such as adrenaline are found associated with invertebrate chromaffin tissue and 5-hydroxytryptamine is also found in many invertebrate nervous syems. It is tempting to speculate that these substances are performing some function in invertebrates and often their experimental administration does elicit a response. However, although it seems likely that chemical transmission does play a part in invertebrate function, very little evidence has so far been obtained to support this view: it is clearly a potentially fruitful field for future research, although the technical problems are daunting.

3 Regulation through Nervous Pathways

This chapter will be principally concerned with a description of methods used to regulate mammalian blood pressure and the mechanisms which ensure the stability of the gas tensions and pH of mammalian blood. The relative constancy of these factors in the face of a variety of physiological and pathological conditions is vital to efficient function. Extensive investigation has revealed the existence of extremely complex and sophisticated homeostatic mechanisms which depend almost entirely upon neuromuscular activity, since the endocrine system plays relatively little part. The properties of comparable regulatory mechanisms in sub-mammalian species will also be discussed.

3.1 Regulation of arterial blood pressure in mammals

Every organ in the body requires an adequate supply of blood in order to function efficiently. Furthermore, an increase in the activity of every organ must be accompanied by an appropriate increase in its blood flow. The blood flow (F) will be related to the hydrostatic pressure difference between the arteries and veins on either side of the organ (ΔP) and the frictional resistance to flow (R) afforded by the small blood vessels within it; such that

$$F = \frac{\Delta P}{R} \tag{1}$$

This equation is, of course, analogous to Ohm's law for the relation between voltage (E), current (I) and resistance (R) in an electrical circuit:

$$I = \frac{E}{R} \tag{2}$$

Most organs can themselves determine their blood flow to some degree, since the metabolic events associated with muscle contraction or secretory activity result in local changes such as decreased pH and partial pressure of oxygen (p.O_2), increased p.CO_2, and the production of metabolites, which cause neighbouring blood vessels to dilate. In addition, in many cases, the nerves which initiate the increased activity in the organ, themselves act on the small blood vessels to cause vasodilation.

Notwithstanding the local vasodilation associated with activity described above, it is clear from equation (1) that the arteriovenous pressure difference must be maintained in order to support the increase in blood flow. In practice, the venous pressure remains relatively stable so that the crucial factor becomes the regulation of arterial pressure. This is of particular importance in the case of the brain which is uniquely dependent upon arterial pressure, since it has relatively little local control of its circulation. If the mean arterial blood pressure falls markedly below its normal value (about 100 mm Hg in man), brain function is impaired and the subject first loses consciousness, then vital functions become progressively slowed until finally respiration ceases and death ensues.

In order to analyse the mechanisms responsible for controlling arterial blood pressure, it is necessary first to define the variables in the system. We shall consider only the systemic circulation (that supplied by the left ventricle) as the pulmonary circulation (the lungs, supplied by the right ventricle) is secondary and has little control of its own.

If equation (1) is applied to the systemic circulation as a whole, neglecting the venous pressure we find that:

$$\text{Flow (cardiac output)} = \frac{\text{Arterial pressure}}{\text{Total peripheral resistance}}$$

The variables in the system controlling arterial pressure are thus the output of the left ventricle and the frictional resistance offered by all the blood vessels in the systemic circulation: it is assumed that the viscosity and volume of blood remain constant.

The cardiac output is of course limited by the return of blood to the heart (no pump can discharge more fluid than it receives), but we shall simplify the system and ignore this factor. The output is the product of heart rate and stroke volume (the volume of blood expelled at each beat). These factors are interrelated since, as the rate increases, cardiac filling and thus stroke volume tend to decrease. Furthermore, stroke volume depends on the vigour of contraction which is in turn related to many factors such as the degree of filling, autonomic activity and hormones from the adrenal medulla. Once again, having briefly mentioned these complexities we shall ignore them and consider cardiac output as a single entity. As such, it is increased by activity in the cardiac sympathetic nerves and decreased by activity in the vagus (parasympathetic) nerves.

The second factor determining arterial blood pressure is the peripheral resistance of the systemic vessels. Among these the principal variable is the diameter of the so-called precapillary resistance vessels; the small arteries and arterioles. These have a thick coat of smooth muscle and are supplied by sympathetic vasoconstrictor fibres. It is variations in the tonic activity of these nerves (see Fig. 2–2) which affords the means of altering the peripheral resistance.

So much then for the effector side of blood pressure regulation; appropriate variation of cardiac output and peripheral resistance can induce very considerable alterations in arterial blood pressure. However, such effector responses must be co-ordinated, which implies first, that there should be some means of monitoring blood pressure and, secondly, that such sensory information should be transmitted to a region where it can be integrated with the animal's requirements and converted into the appropriate executive commands to the heart and blood vessels.

Blood pressure is monitored by sensitive stretch receptors found in the walls of the upper aorta and in the carotid sinuses (Fig. 3–1.a). These receptors transmit impulses up afferent nerves to the brain, and the frequency of discharge is related to the arterial pressure as shown in Fig. 3–1b. Information is transmitted by the 'buffer nerves' from these 'baroreceptors' to special centres in the medulla concerned with cardiovascular control.

There are basically three centres in the brain whose principal concern is the circulation. The vasomotor centre, as its name suggests, is mainly concerned with alterations in sympathetic vasomotor activity but it also controls the sympathetic nerves to the heart. It is subdivided into the 'depressor' and the 'pressor' centres. The first helps to *depress* the blood pressure both by decreasing sympathetic stimulation of the heart and thereby reducing cardiac output, and also by decreasing activity in sympathetic vasoconstrictor nerve fibres and thus causing vasodilation and decrease in peripheral resistance. The pressor centre has exactly the opposite effects to the depressor centre and thus *increases* blood pressure by increasing cardiac output and the peripheral resistance. The third centre is the cardioinhibitory centre and it is associated simply with the vagus nerves to the heart; when active it decreases cardiac output and thus combines with the depressor centre to lower the blood pressure.

Figure 3–2 summarizes the way in which baroreceptors, buffer nerves, medullary centres, efferent nerves and the effector mechanisms combine to stabilise blood pressure. A fall in pressure decreases baroreceptor discharge thereby stimulating the pressor centre by decreasing its inhibition while inhibiting the depressor and cardioinhibitory centres by decreasing their stimulation. The resulting increase in cardiac output and peripheral resistance restores normal pressure. Conversely, a rise in pressure inhibits the pressor centre while stimulating the depressor and cardioinhibitory centres; the cardiac output and peripheral resistance fall and the blood pressure declines to resting levels.

3.1.1   Regulation of blood pressure in submammalian species

The control of the arterial blood pressure in birds, reptiles and amphibia is essentially similar to that described previously for the mammal. Thus, the vagus nerve depresses the heart while sympathetic activity stimulates it and there are baroreceptors in the arterial system with afferent fibres running in the vagus and glossopharyngeal nerves. There are, of course, certain deviations from the general pattern such as a sympathetic innervation via the vagus nerve to the heart in some amphibia, but the basis for reflex control of blood pressure has been for the most part established at the level of the amphibia. In jawed fish there

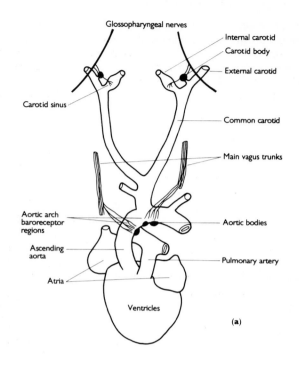

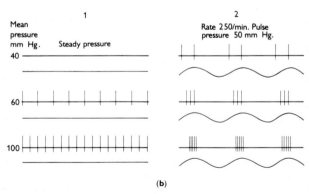

Fig. 3–1 (a) Diagram to show the anatomical location of the arterial barorecptors and their afferent nerves, (b) Discharge characteristics of single fibre from a carotid sinus baroreceptor in response to (1) maintained pressure, (2) pulsatile pressure. (After GREEN, J. H. (1958) in *Reflexogenic Areas of the Cardiovascular System*, HEYMANS and NEIL, eds. Churchill Livingstone, Edinburgh.)

is no sympathetic nerve supply to the heart, although in some species the vagus nerve contains some adrenergic fibres which may serve to increase heart rate. Vagal inhibition is pronounced, however, and is modulated by afferent impulses in the branchial nerves from baroreceptors in the gill arteries which respond to increased intravascular pressure: this is probably the precursor of the depressor reflex in higher forms while the

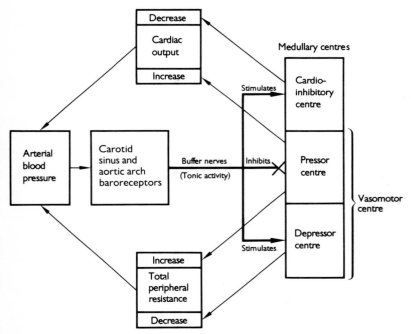

Fig. 3–2   Action of the arterial baroreceptors on the medullary cardiovascular control centres.

pseudobranch vessels are homologous with the carotid sinus and carotid bodies of birds and mammals. In some fish, the heart can also be slowed reflexly in response to stimulation of the skin or if the flow of water across the gills ceases: the adaptive value of such responses is obscure. In certain more primitive species such as the hagfish there is no innervation of the heart. Little is known about vasomotor control in fish: most peripheral blood vessels have a sympathetic innervation but some of these nerves may in fact release ACh. ACh is a vasodilator in mammals, but tends, in contrast, to have a vasoconstrictor action in fish and serves to help increase the blood pressure by raising the peripheral resistance. In fish, adrenaline also causes constriction in systemic vessels, but dilates the branchial vessels. As a first approximation from the available evidence, it appears that, in fish, blood pressure is lowered principally by vagal depression of cardiac output and raised by sympathetic constriction of peripheral vessels. This contrasts with the situation in mammals where *both* cardiac output *and* the peripheral resistance can be altered in either direction by the autonomic nervous system.

In invertebrates little is known about the regulation of pressure and flow in the circulating body fluids. The most advanced arrangement is that of the cephalopod molluscs which have a closed system with a powerful systemic heart: in octopus, pressures as high as 60 mm Hg have been recorded. Molluscan hearts appear to be principally dependent upon the fluid pressure within them; if well distended they beat maximally whereas if the pressure is low they stop. Cephalopod hearts are richly innervated by both inhibitory and excitatory nerves, but little is known of the physiological role played by these nerves in the control of the heart.

Most invertebrates have an open circulatory system where pressure and flow is low and variable while the haemolymph volume is large. Flow in such systems is often more dependent upon somatic muscle contraction than on the action of the heart as such: an arrangement which automatically ensures a brisker circulation during activity. In most crustacea, the rhythmic contractions of the heart originate in a group of nerve cells associated with the heart: the rhythm is thus neurogenic. Rate can be altered by both inhibitory and accelerator nerves coming from the suboesophageal ganglion. The principal control of heart rate, as in the cephalopods, seems to be the distending pressure, but in crustacea, inflation of the heart probably increases the rate via the stretch imparted to the dendrites of the pacemaker nerve cells on the surface of the heart. Reflex alterations in heart rate can be evoked by stimulation of the body surface; pain and chemical stimulation are particularly effective. In addition, a hormone, which may be 6-hydroxytryptamine, secreted by the pericardial gland causes a marked increase in heart rate. The factors controlling its release and its physiological importance relative to the nervous influences on the heart remain unclear.

The enormous diversity of insects, together with the fact that the heart does not play a part in respiratory transport, gives rise to much variation in the form and function of the hearts found in the group. The beat may be myogenic or neurogenic. The heart may beat only if its dorsal suspensory ligaments are intact and it may or may not be supplied with inhibitory or acceleratory nerves from the central nervous system. Virtually nothing is known about normal cardiovascular control in insects, although it has been demonstrated that the heart of the sphinx moth, which is in the abdomen, can be influenced by changes in thoracic temperature: an effect presumed to be mediated via the nerves to the heart.

One final point about circulatory fluid pressure in the invertebrates is that changes in pressure may also serve a hydraulic function. The use of internal fluid pressure for skeletal or locomotor functions is common in many invertebrate groups and clearly, under these circumstances, pressure differences directed toward the circulation of the body fluid assume secondary importance.

3.2   Regulation of the concentration of O_2 and CO_2 in mammalian blood

Cellular respiration requires O_2 and produces CO_2 and, although certain tissues such as skeletal muscle can function anaerobically for short periods, it is a general prerequisite for efficient function that tissues should be adequately supplied with O_2 and that excess CO_2 should be rapidly removed. In mammals, blood with its respiratory pigment haemoglobin and its intricate buffer systems, acts as the vehicle to convey O_2 to peripheral tisues and also carries away CO_2. Venous blood from the systemic circulation is therefore low in O_2 and contains relatively large quantities of CO_2: it returns to the right heart and is pumped through the lungs. There the gases equilibrate, by diffusion through the delicate alveolar walls, with the air contained in the alveolar spaces, thereby recharging the blood with O_2 and removing excess CO_2. This pulmonary gas exchange depends upon the presence of a gas mixture in the alveoli which contains precisely the correct quantities of O_2 and CO_2. Equilibration between alveolar air and the blood in the pulmonary capillaries is so rapid that the gas tensions in pulmonary venous blood (equivalent to systemic arterial blood) are virtually identical with those in alveolar air. What we must consider therefore is the problem of how the composition of alveolar air is maintained in the face of wide variations in the O_2 requirement and CO_2 production of the body.

3.2.1   Alveolar ventilation

The mammalian lung consists essentially of a complex system of blind-ended tubes only the terminal parts of which (alveoli) possess the delicate membrane necessary to permit rapid diffusion of respiratory gases: there are some 300 million alveoli in the human lungs representing a total area for gas exchange of about 70 m². This area can support an O_2 transfer in excess of 5000 ml min⁻¹ during maximal exercise and caters easily for the resting O_2 uptake of about 250 ml min⁻¹.

The alveoli are ventilated by a tidal movement of air as the lungs rhythmically inflate and deflate as a result of the increase and decrease in the volume of the thorax produced by the movements of the diaphragm and muscles in the chest wall.

Unlike the intrinsic rhythmicity of the heart which persists even when its nerve supply is cut, the respiratory rhythm originates in the respiratory centres of the brain. The rate of breathing and the depth of each breath (tidal volume) therefore depend on the activity of these centres which thus ultimately govern the composition of alveolar air and therefore the partial pressures of O_2 and CO_2 in arterial blood.

The question we must answer then is how the respiratory centres in the brain regulate the pattern of respiration so that the alveolar gas mixture and thus the arterial O_2 and CO_2 remain relatively constant in the face of up to twenty-fold increases in the demand for oxygen or the production of CO_2. As with all examples of homeostasis, the question resolves itself into two parts; the identity and measurement of the input and how such sensory information is used to promote an appropriate corrective output. In the case of blood pressure regulation discussed earlier in this chapter, the input measured was obvious and was clearly monitored in the distinct baroreceptive parts of the arterial tree. In the case of respiration the situation is more complex for there are at least two candidates for the role of controlling variable (blood p.O_2 and p.CO_2) and a number of sites both peripheral and in the brain where these variables are monitored.

The resting subject at sea level   In order to decide whether the principal factor which controls respiration in a resting subject is arterial p.O_2 or arterial p.CO_2 it is necessary first to approach the problem analytically. Thus we must eliminate *one* of the variables and determine the effect on respiration of altering the other. The results of such experiments are shown in Fig. 3–3. In a subject made to breathe air containing progressively less O_2 (atmospheric air contains approximately 20% O_2) there is little increase in respiratory minute volume (respiratory rate × tidal volume) until the O_2 content of the inspired air is reduced to half its normal value. At this point the p.O_2 of alveolar air and thus arterial blood has fallen to 40 mm Hg (normal alveolar p.O_2 breathing atmospheric air at sea level is about 100 mm Hg). As the O_2 concentration of the inspired air is reduced further, respiration increases gradually. In fact, it is necessary to reduce the level to as low as 5% before the minute volume achieves double the resting value. Hypoxia (lowering of the arterial p.O_2) would therefore seem to be an extremely ineffective stimulus for increasing respiration. Compare the response to hypoxia with the converse experiment shown in Fig. 3–3 where the percentage of CO_2 in inspired air was progressively increased. In this situation, the respiratory minute volume increases almost immediately the arterial p.CO_2 rises above the normal value of 40 mm Hg. These experiments would seem to indicate that the alveolar (arterial) p.CO_2 is the principal controlling factor in the regulation of respiration in a resting subject at sea level; thus, any tendency for the alveolar p.CO_2 to increase, results in an immediate increase in respiration. No such sensitive mechanism operates for oxygen: the percentage of O_2 in inspired air can be halved before any significant increase in respiration is produced.

In physiology, however, one must be aware of the pitfalls of the

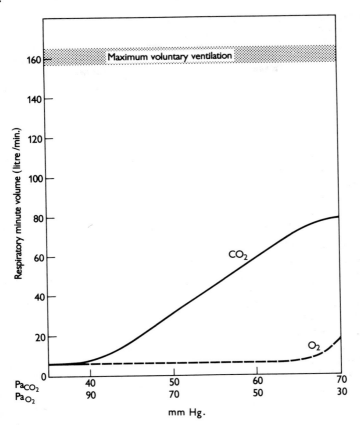

Fig. 3-3 Acute respiratory response to decrease in inspired O_2 concentration *or* increase in inspired CO_2 concentration in a 70 kg human male subject. Abscissae, arterial $p.O_2$ and arterial $p.CO_2$: these relate to the lines labelled O_2 and CO_2 respectively. Responses are compared with the maximum voluntary respiratory effort the subject could produce. Ordinate; respiratory minute volume. (Modified from RUCH, T. C. and PATTON, H. D. (1974), *Physiology and Biophysics* II, 20th Edn., W. B. Saunders Company, Philadelphia and London.)

analytical approach to problems. The experiments cited above might lead one to believe that respiratory regulation devolved simply upon the maintenance of a stable arterial $p.CO_2$ since *independent* variation of alveolar $p.CO_2$ and $p.O_2$ produced such a clear-cut difference between the sensitivity of the resulting respiratory responses. However, in life, in contrast to these experiments, the two variables are not normally independent of each other because inadequate ventilation of the lungs results in both accumulation of CO_2 *and* a deficiency of oxygen (asphyxia). What then of the interactions of these two factors? Figure 3–4 shows that the respiratory response to increased alveolar p.CO_2 is greatly enhanced when it is accompanied by a decrease in alveolar p.O_2. Oxygen deficit thus *sensitizes* the response to increased p.CO_2.

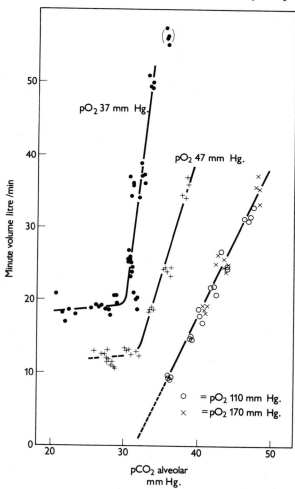

Fig. 3–4   Effect of hypoxia on the ventilatory response to CO_2 inhalation in a human subject. Inspired O_2 was adjusted to maintain the alveolar p.O_2 values indicated. (Modified from NIELSEN, M. and SMITH, H. (1951), *Acta Physiol. Scand.*, **24**, 293.)

Effects of altitude   Ascent to high altitude is accompanied by progressive hypoxia since, although the composition of atmospheric air and thus the percentage of O_2 remains constant, the $p.O_2$ decreases simply because the total barometric pressure declines: under these circumstances the stimulation of respiration by hypoxia becomes of paramount importance. Above about 3000–3500 m respiration rate is determined primarily by the need to obtain oxygen rather than, as at lower altitudes, the need to eliminate CO_2.

3.2.2   Measurement of $p.O_2$ and $p.CO_2$ in arterial blood

We have seen how the rhythm of respiration originates in the brain and that this rhythm is normally modulated appropriately to maintain the alveolar $p.CO_2$ constant, although under certain circumstances such as high altitude the alveolar $p.O_2$ may also play a part in the regulation. How does the brain know what the alveolar $p.O_2$ and $p.CO_2$ values are, since as far as is known, there are no chemoreceptors in the lungs? In fact the gas tensions are monitored in arterial blood which is in virtual equilibrium with alveolar air. The $p.CO_2$ is monitored by the brain directly: receptors just above the ventral surface of the medulla respond to changes in the pH of the cerebrospinal fluid which accurately reflects arterial $p.CO_2$. A decrease in $p.O_2$ on the other hand is detected by the peripheral chemoreceptor cells which are located in the carotid and aortic bodies (see Fig. 3–1). These cells respond to decreased arterial $p.O_2$ and to a lesser extent to increased $p.CO_2$ or decreased pH and send impulses up sensory nerve fibres to the respiratory centres.

The respiratory response to increased $p.CO_2$ depends almost exclusively upon the chemoreceptors in the brain stem: it is little affected by cutting the sensory nerves from the carotid and aortic bodies. Conversely, the response to decreased $p.O_2$ is entirely dependent upon the peripheral chemoreceptors: if the sensory nerves are cut, the only effect of hypoxia is to reduce respiration by a direct depressant effect in the respiratory centres.

3.2.3   Overall control of respiration in mammals

The overall control of respiration is summarized in Fig. 3–5 from which it will be seen that in addition to the $p.CO_2$ and $p.O_2$ a number of other factors may have subsidiary effects. Of these, the pH of arterial blood is of considerable importance since a decrease below the normal value of 7.4 (acidosis) promotes an increase in respiration via an effect both on the brain and the peripheral chemoreceptors. Accumulation of CO_2 in the blood is associated with acidosis, as occurs if respiration is inadequate (respiratory acidosis).

$$CO_2 + H_2O \rightleftharpoons H_2CO_3 \rightleftharpoons H^+ + HCO_3^- \tag{3}$$

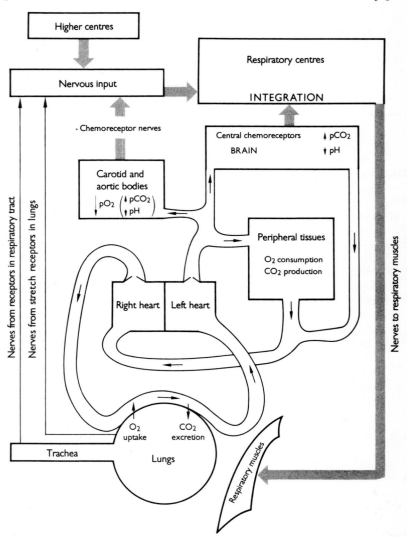

Fig. 3–5 Interaction of nervous and chemical influences in the regulation of respiration (see text).

Conversely, it will also be seen from equation (1) that addition of hydrogen ions to the blood such as occurs when lactic acid enters the blood from active muscle will result in the production of CO_2 (metabolic acidosis). The pH and pCO_2 are therefore intimately interrelated in the control of respiration. Figure 3–5 also illustrates a number of non-chemical factors known to influence respiration such as the higher centres of the brain (voluntary alteration in respiration, speech, thermal panting, etc.) receptors in the lungs regulating the degree of inflation, receptors in the trachea and bronchioles (coughing, sneezing) and receptors in joints which, together with acid produced by muscle, changes in gas tensions and information from higher centres, are responsible for the increased respiration during exercise.

3.2.4   *Regulation of respiration in submammalian species*

The consumption of O_2 and production of CO_2 by animal tissues has necessitated the development of specialized apparatus to facilitate gas exchange with the external environment. The apparatus developed by different animals is, however, so diverse that space will not permit description and will only allow consideration of the *control* of the respiratory apparatus in a few representative examples. Two preliminary generalizations are worth making. First, from the annelids upwards, animals depend upon a pacemaker of some sort to provide the basic respiratory rhythm and such factors as temperature and mechanical and chemical stimuli act by influencing this pacemaker. Secondly, the major chemical determinant for respiratory control is CO_2 in the case of terrestrial animals and O_2 in the case of aquatic animals.

In birds, like mammals, CO_2 is more important than O_2 in regulating minute volume, principally by direct action on the medullary respiratory centre, although there are additional CO_2 receptors actually in the lungs and upper respiratory passages. Respiratory changes in response to hypoxia are slight although chemoreceptors corresponding with the mammalian carotid bodies are present. Carotid and aortic bodies as such are not found in amphibia or fish and although there are homologous structures in reptiles they are more sensitive to CO_2 than to hypoxia. Fish are much more sensitive to hypoxia than to build-up of CO_2, although changes in the levels of either gas in the water passing the gills will cause respiratory changes. The location of the O_2 and CO_2 receptors is not known since reduced responses persist when the glossopharyngeal and vagus nerves are cut: there are probably both peripheral and central chemosensitive areas. It is of interest that the electric eel, which is an obligatory air-breather, is more sensitive to CO_2 than to hypoxia. In some species of fish, a profound decrease in the O_2 content of the water can trigger off an emergency respiratory response. Thus, some South American swamp dwelling fish can gulp air into the swim bladder and use O_2 from this source while certain of the loaches (*Cobitidae*) change from branchial respiration to intestinal breathing of air. As well as the behavioural changes above, many teleosts can modify the efficiency of gaseous exchange by vascular responses in the gill filaments. There appear to be alternative pathways for blood; either via the respiratory vessels in the gill lamellae or via non-respiratory vessels in the centre of the gill filament. The distribution of blood probably depends on nervous and endocrine factors and experimentally it has been found that adrenaline promotes blood flow through the lamellae and ACh directs flow through the filament centre.

In invertebrates, there are many examples of the way in which aquatic respiration is stimulated by decrease in the O_2 content of the water while air-breathing relatives have become more sensitive to CO_2 accumulation. Terrestrial crustacea such as *Ligia oceanica* (the shore slater) show little increase in respiration even when the $p.O_2$ is halved, whereas aquatic species, e.g. *Gammarus pulex* (the fresh-water shrimp) accelerate respiratory movements in response to a relatively small fall in dissolved O_2. Likewise among the molluscs, terrestrial pulmonates such as *Helix* open the pneumostome in response to increase in $p.CO_2$ and do not close it until normal $p.CO_2$ is restored. Aquatic pulmonates obtain O_2 through the skin as well as from the mantle lung and are relatively insensitive to increased $p.CO_2$. Most bivalves react to low O_2 tensions by pumping water more rapidly over the gills, but sufficiently high levels of CO_2 can actually depress respiratory activity, causing the shell to close and the cilia to cease beating.

In the smaller and less active insects, and in the larval and pupal stages, diffusion of air through the tracheal system is often sufficient for adequate gas exchange, but more dynamic species show active ventilation of the system together with control over opening the spiracles. In insects, unlike mammals, expiration is active and results from movements of the body wall such as dorso-ventral flattening of the abdomen which forces air from the tracheae. Segmental ganglia are responsible for the rhythmic ventilatory movements, but these ganglia are affected by secondary centres in the thorax which are sensitive to the $p.CO_2$. Moderate increases in $p.CO_2$ thus result in increased ventilation but the gas becomes anaesthetic in high concentration. The opening of the spiracles, in the cockroach at least, seems to be sensitive to both O_2 and CO_2: this is probably a direct action on the spiracles. Increase in $p.CO_2$ may, in some species, also help to synchronize spiracular opening with breathing movements. Hypoxia increases the sensitivity of the spiracles to CO_2, conversely, high O_2 will raise the CO_2 threshold. Receptors sensitive to CO_2 are found in the central ganglia, the spiracles and, in some insects such as honeybees, there are additional CO_2 receptors on the antennae. Finally, the fact that hive bees are stimulated to ventilate the hive by low O_2 or high CO_2 is perhaps an example of communal homeostasis!

4 Regulation through Endocrine Pathways

4.1 The regulation of blood glucose concentration in mammals

The title above is perhaps a little misleading with regard to the contents of this chapter, because blood glucose concentration represents merely the tip of a very large iceberg: the overall problem of ensuring an adequate supply of metabolic energy to the cells of the body. All living cells require a continuous supply of metabolic energy; furthermore, this requirement fluctuates in step with the functional activity of the cell. In mammals, energy is supplied to the cell in the form of glucose or various fat derivatives carried to it by the blood. These energy substrates enter the blood from storage reservoirs or by intestinal absorption of components of the diet, although it should be remembered that even energy from storage reservoirs originates from the diet.

The problem of how dietary components are distributed to the various tissues is exceedingly complex and space will permit only a relatively superficial examination. Nevertheless, since it represents the most elaborate endocrine regulated homeostatic system known, we will devote an entire chapter to its consideration.

Figure 4–1 depicts a grossly oversimplified hydraulic model of glucose circulating in the blood (the blood glucose pool) and its relationship to the glycogen stores of the body. The blood glucose concentration is represented as the level of fluid in a tank: the 'blood glucose pool'. The level of fluid will clearly depend upon the algebraic sum of the inflow and outflow of glucose. There are two sources of inflow: first, the intermittent addition of glucose to the blood which occurs after meals as the result of the absorption of the carbohydrate in the diet and, secondly, the release of glucose from glycogen stores, principally that in the liver (glycogenolysis). Glucose is lost from the pool as it is utilized by tissues. It should be noted that most tissues can metabolize either glucose or fat derivatives but some tissues, notably the brain, depend exclusively upon glucose for their metabolic energy. Glucose can also be removed from the blood and stored as glycogen (glycogenesis) and, if the concentration becomes sufficiently high, it can be lost into the urine. Normally in man blood glucose concentration remains within the range indicated by the shaded area on Fig. 4–1. Thus, during the absorptive phase after a meal, levels may reach 6·1 to 6·7 mmol l^{-1} (110–120 mg glucose 100 ml^{-1} blood), whereas after a moderate fast the concentration may fall to 3·5–5·0 mmol l^{-1} (70–80 mg 100 ml^{-1}). At the upper end of the normal range there is net glycogenesis, i.e. more glucose is being incorporated

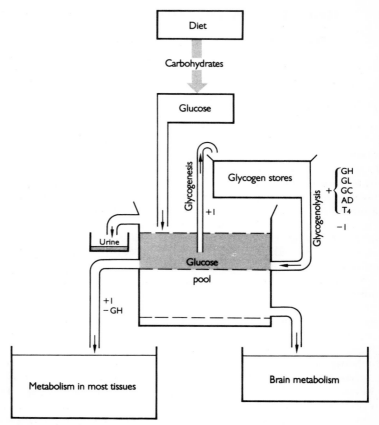

Fig. 4–1 Diagram to illustrate the concept of the blood glucose 'pool' and the processes which influence the size of the 'pool'. Hormones which stimulate (+) or inhibit (−) particular processes are abbreviated thus: I, insulin: GH, growth hormone: GL, glucagon: GC, glucocorticoids: AD, adrenaline: T_4, thyroxine.

into glycogen stores than is being released. If the blood glucose concentration falls below about 3·3 mmol l^{-1} (60 mg 100 ml^{-1}) no glucose can enter cells, other than those of the brain, so that such cells must depend upon fat for their metabolic energy. If the concentration continues to fall (profound hypoglycaemia) insufficient glucose is available even for the brain and convulsions, coma and death ensue.

How then are the movements of glucose directed? The answer lies principally in the actions of various hormones which serve as 'flow valves' and regulate appropriately the entry of glucose into cells, glycogenesis and glycogenolysis. Six hormones exert major effects which are summarized in Figs 4–1 and 4–2 where + indicates stimulation of a process and − inhibition. Consider first the entry of glucose into cells; this process in most tissues other than the brain, gut and kidney depends upon the presence of insulin. The secretion of insulin is primarily determined by the blood glucose concentration (p. 14) such that insulin is not secreted if blood glucose concentration is less than 3·3 mmol l^{-1} (60 mg 100 ml^{-1}) and progressively more insulin is secreted as glucose concentration increases above this threshold. The consequence of this is that in times of glucose deficit (<3·3 mol l^{-1}: 60 mg 100 ml^{-1} blood), the little glucose available cannot enter cells whose glucose entry is insulin dependent and it is thus reserved for tissues such as the brain where entry does not depend on insulin and which furthermore have an absolute requirement for glucose. Under these circumstances, cells which get no glucose rely on fat for their metabolic energy. In addition to its effect on glucose entry into cells, insulin stimulates glycogenesis and inhibits glycogenolysis, both of which actions are also directed towards lowering blood glucose concentration. It is clear then that insulin, the secretion of which is augmented by increased blood glucose, serves by at least three means to lower blood glucose (hypoglycaemic actions) – negative feedback. Insulin is of vital importance to the regulation of metabolic energy supply for it is the *only* hormone with hypoglycaemic actions. As can be seen from Fig. 4–1, glucagon, glucocorticoids, adrenaline, growth hormone and thyroxine all stimulate the breakdown of glycogen, while growth hormone slows glucose entry into certain cells. These hormones therefore tend to raise blood glucose concentration (hyperglycaemic effects) and, as might be expected, their secretion is stimulated by a fall in blood glucose. There is therefore a dual negative feedback control imposed upon blood glucose concentration: hyperglycaemia stimulates insulin secretion, which promotes a hypoglycaemic response, and conversely, hypoglycaemia inhibits insulin secretion while stimulating the release of other hormones with hyperglycaemic actions.

4.2   Interrelations between carbohydrate, fat and protein metabolism

It would be extremely naïve to consider the regulation of blood glucose concentration in isolation, as it is so intimately involved with the metabolism of both fat and carbohydrate. Figure 4–2 attempts to illustrate some of the principal linking pathways.

Fat in the diet is absorbed into the circulation either via the lymphatics as triglycerides which pass direct into the fat depots or, after hydrolysis in the gut, as glycerol and free fatty acids (FFA). The FFA pool in the blood comprises FFA absorbed from the gut and FFA released by hydrolysis of triglycerides stored in adipose tissue. The fat stores of the body constitute by far the most extensive reserve of metabolic energy. It has been

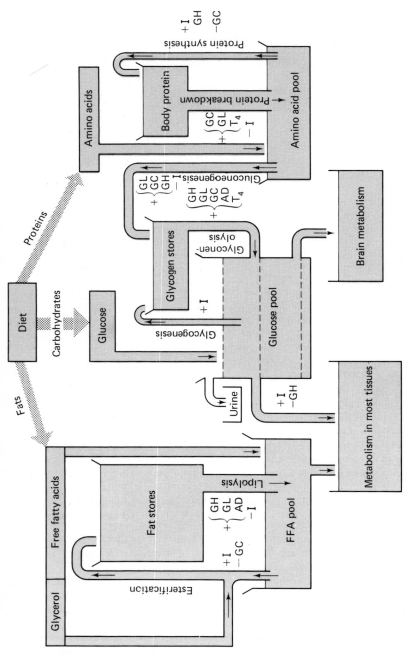

Fig. 4–2 General interactions between carbohydrate, fat and protein within the body. Hormones which stimulate (+) or inhibit (−) particular processes abbreviated thus: I, insulin: GH, growth hormone: GL, glucagon: GC, glucocorticoids: AD, adrenaline; T_4, thyroxine.

estimated that in man $c.$ 420 000 kJ (100 000 kcal) are stored as fat, $c.$ 3000 kJ (700 kcal) as carbohydrate and $c.$ 105 000 kJ (25 000 kcal) as protein. The fat stores can be broken down when required by lipolytic enzymes to release FFA into the blood for cellular metabolism. Conversely, in times of plenty, such as after a meal, plasma FFA and glycerol are esterified to replenish the stores. As discussed previously in the case of carbohydrate mobilization, the disposition of fat reserves is again controlled by endocrine 'flow-valves', which act by regulating the function of specific enzymes. Figure 4–2 shows that insulin favours the formation of stored fat while inhibiting lipolysis: growth hormone, glucagon and adrenaline promote lipolysis.

Protein is absorbed from the intestine as amino acids which enter the blood 'pool' from which they can either be taken up by cells and incorporated into protein or converted, via such intermediates as pyruvic acid, into glucose: this latter process is called gluconeogenesis and occurs principally in the liver. Once again, the various pathways are regulated by hormone-controlled 'flow valves' as shown in Fig. 4–2.

It is clear that there is an extremely sophisticated regime of hormones available to mammals to ensure an optimum supply of metabolic energy for the cells of the body. In order to see how the system works in concert we shall examine the everyday responses to the feeding-fasting cycle.

4.3   Effect of feeding

Figure 4–3 summarizes the endocrine responses to feeding, the key to the understanding of which is appreciation of the overwhelming effect of the insulin released. There are a number of factors which promote the secretion of insulin. These are now tabulated, numbered as in Fig. 4–3.

(1) Digestion is associated with the secretion of secretin, pancreozymin and a glucagon-like substance from the duodenum: all these hormones stimulate insulin secretion.

(2) Activity in the vagus nerve during digestion is thought to cause insulin release in some species.

(3) As the products of digestion are absorbed into the blood the increases in (a) glucose concentration and (b) amino acid concentration both stimulate insulin secretion.

(4) Reference to Fig. 4–3 will show that insulin promotes
  (a) Glucose uptake and utilization by cells
  (b) Glycogenesis
  (c) Protein synthesis
  (d) Esterification

(5) It should be noted that the increased blood glucose concentration tends to depress the secretion of such hormones as glucagon, glucocorticoids, growth hormone and adrenaline.

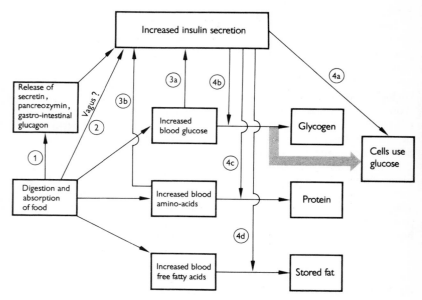

Fig. 4–3 Endocrine and metabolic changes following feeding (see text).

Insulin is sometimes called the 'hormone of plenty'. Its copious secretion after meals and the effective way in which it prevents excessive increases in glucose, amino acid and fatty acid concentrations in the blood by promoting their storage at this time, show that this title is amply justified.

4·4 Effect of fasting

Figure 4–4 shows the endocrine status after a moderate fast, as in a human subject immediately before breakfast. Once again the figures tabulated below correspond with those in Fig. 4–4. In this situation the crucial factor is the low blood glucose concentration (hypoglycaemia):

(1) Hypoglycaemia depresses insulin secretion.
(2) In the absence of insulin, glucose uptake by most cells is inhibited.
(3) The brain does not require insulin for glucose uptake and therefore has sole call on available glucose.
(4) Hypoglycaemia stimulates secretion of glucagon, glucocorticoids, growth hormone and, if severe, adrenaline.
(5) These hormones stimulate glycogenolysis to maintain entry of glucose into the blood.
(6) They stimulate protein breakdown to increase the amino acid pool.

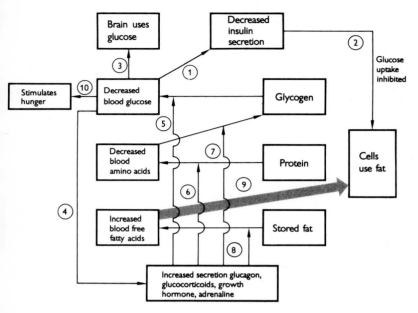

Fig. 4–4   Endocrine and metabolic changes during fasting (see text).

⑦ They promote gluconeogenesis.

⑧ Perhaps most important they cause breakdown of stored fat.

⑨ The FFA thus released becomes available to maintain the function of those cells deprived of glucose.

⑩ Hypoglycaemia stimulates hunger and thus induces the animal to supplement its depleted energy reserves (see Chapter 5).

These two examples demonstrate the complex way in which the secretion of many hormones may be integrated to ensure that the supply and demand of cellular metabolites maintains a homeostatic balance.

4.5   Control of metabolism in submammalian species

We shall not dwell on the control of metabolism in other vertebrate groups for, as outlined in Chapter 2, the hormones involved are comparable with those in mammals and the general strategy of the regulatory mechanisms is closely similar. Moreover, as relatively little is known of the regulatory pathways in invertebrate groups other than insects, our discussion will be restricted to flying insects.

It was established in 1956 that the principal sugar in insect haemolymph is trehalose, a symmetrical disaccharide of glucose (α-Dglucopyranosyl-α-D glucopyranoside): in some species this sugar can be found in such high concentrations in haemolymph as to be almost in crystalline form. Like blood glucose in vertebrates, haemolymph trehalose comprises the mobile energy 'currency' in insects, as shown in Fig. 4–5.

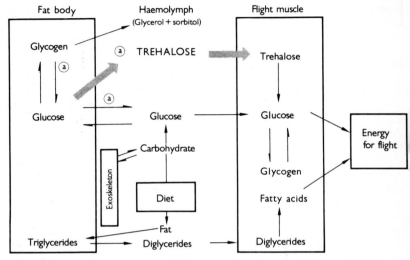

Fig. 4–5 Energy stores and the provision of energy for insect flight.

The major energy storage organ in insects, equivalent to the vertebrate liver, is the fat body. Here, glycogen, derived mainly from dietary carbohydrate, composes the principal energy reserve, but the fat body also stores triglycerides from the fat in the diet. Trehalose is synthesized within the fat body from two hexoses derived from glycogen and passes into the haemolymph. It is then available to peripheral tissues, particularly the flight muscles, which are easily the greatest users of energy in the insect body. Once within the flight muscles, trehalose is broken down by the enzyme trehalase to form two molecules of glucose: glucose also comes from the limited glycogen stores within the muscle. An alternative source of energy for flight is fat which is released from the triglyceride stores in the fat body into the haemolymph as diglycerides. The diglycerides then enter the flight muscles where they are hydrolysed and oxidized. In fact, insect flight muscles manifest the most active fat catabolism yet described, exceeding even that seen in homeothermic animals. The relative importance of glucose or fat as fuel for the flight muscles is different in the various groups of flying insects. Thus, most dipterans and hymenopterans use exclusively carbohydrate while all lepidopterans depend entirely on fat. Orthopterans, homopterans and coleopterans use a combination of fat and carbohydrate for flight initiation, but then, having depleted the fat body carbohydrate and blood trehalose virtually completely, subsequently rely upon fat.

Relatively little is known of the control of the metabolic pathways described above and illustrated in Fig. 4–5. The synthesis of trehalose within the fat body is subject to a kind of direct negative feedback: if trehalose concentration in the haemolymph is low, hexoses within the fat body are combined under the action of a specific enzyme to form trehalose. However, trehalose itself inhibits the activity of this enzyme, so, at high trehalose concentrations further trehalose synthesis is inhibited and free hexoses are incorporated into glycogen instead. The breakdown of fat body glycogen and the consequent formation and release of trehalose and glucose (pathways labelled ⓐ in Fig. 4–5) is promoted by a hormone from the corpus cardiacum. This has been demonstrated in *Periplaneta* (cockroach), but it is not known yet whether it occurs in all species and what stimulates the release of the hormone, although it is tempting to draw a parallel between this hormone and the actions of the various hyperglycaemic hormones in mammals. The release of triglyceride stores from the fat body is also known to be under endocrine control. A hormone which mobilizes fat (the adipokinetic hormone) is secreted from the corpus cardiacum in insects such as the locust and serves to mobilize fat in preparation for prolonged migratory flights.

Three further points concerning insect metabolism merit a brief consideration. First, there is evidence in some species for gluconeogenesis within the fat body; dietary amino acids contribute to glycogen formation. Secondly, fat body glycogen can be converted to polyhydric alcohols such as sorbitol and glycerol: this process is stimulated in an unknown way by exposure to cold and the alcohols produced enhance survival in extreme cold by lowering the supercooling point of the body. In the prepupae of *Bracon cephi* (wheat-stem sawfly parasite) glycerol may comprise 25% of body weight and allows survival down to −40°C. Finally, chitin which occurs bound to protein in the exoskeleton may act as a reserve nutrient. The metabolic mobility of chitin is now the subject of extensive study and it is clear that cuticle polysaccharide is extremely labile and may be resorbed during starvation while, conversely, deposition of endocuticle follows feeding. However, as is currently the case in so much invertebrate physiology, the factors which regulate the metabolic behaviour of chitin remain to be defined.

5 Homeostatic Mechanisms involving both Nerves and Hormones

5.1 The regulation of energy intake

Animals, like physical systems, obey the First Law of Thermodynamics:

Energy intake = work done + heat produced + energy stored

It is clear therefore that the amount of chemical energy contained in an animal's diet must be geared to (*i*) its work output, and (*ii*) its heat production. A diet which exactly fulfils these requirements is called a 'maintenance diet'. If an animal is fed more than its maintenance diet it gains weight, if fed less than its maintenance diet, it loses weight. Most healthy adult humans maintain their body weight remarkably constant over periods of many years, which implies that they possess the ability to relate their food intake accurately to their metabolic requirements.

The mechanisms which control food intake are extremely complex and, even now, are still imperfectly understood. Nevertheless, it is worthwhile to examine the available evidence in order to gain some insight into this elaborate example of homeostatic control, which involves the interaction of both nervous and hormonal control mechanisms.

5.1.1 Hypothalamic centres

Abundant evidence exists to implicate the hypothalamus in the regulation of food intake. Results from damage to the hypothalamus, either by disease in man, or by experimental lesions in animals, and results of electrical stimulation of areas of the hypothalamus in animals have revealed the existence of two opposing control 'centres'. These centres have been named the 'feeding centre' and the 'satiety centre'. The feeding centre in the lateral hypothalamus, as its name suggests, promotes feeding; the satiety centre, in the ventromedial hypothalamus, inhibits the feeding centre. We have therefore, at first sight, a simple negative feedback loop (Fig. 5–1). Damage to the system upsets the feedback (Fig. 5–2); thus, if the satiety centre is destroyed the animal eats excessively (hyperphagia) and becomes grossly obese. Conversely, if the feeding centre is destroyed the animal ceases to eat (aphagia) and, unless fed artificially, slowly dies of starvation. So much is well established – what remains unclear is the nature of the satiety signal.

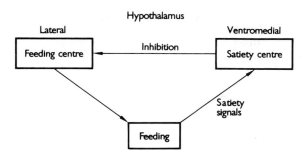

Fig. 5–1 Basic feedback loop in the control of feeding (see text).

5.1.2 Satiety signals

What is it that stimulates the satiety centre to terminate feeding? Three possibilities exist. First, signals from the gastrointestinal tract may be involved, secondly the signal may be thermal and related to body or environmental temperature or, finally, the signal may be a chemical one involved with the absorption of dietary components into the blood.

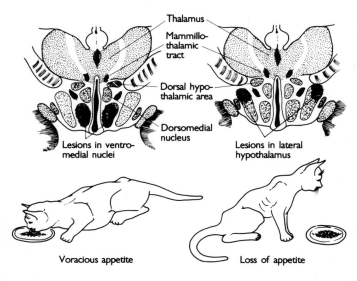

Fig. 5–2 The effect of lesions in the hypothalamus on feeding behaviour. (Adapted, with permission, from *The CIBA Collection of Medical Illustrations* by Frank H. Netter, M.D. © Copyright 1953, 1972 CIBA Pharmaceutical Company, Division of CIBA-GEIGY Corporation. All rights reserved.)

(a) GASTROINTESTINAL SIGNALS After a large meal one is aware of a feeling of fullness or distension, which effectively dissuades one from eating more. Recent evidence suggests, however, that such sensations are of relatively little importance in the control of overall food intake, although clearly they influence the size of individual meals. In rats, dilution of the food with inert materials, such as cellulose, does not disrupt regulation of food intake: the animal will consume its requisite calorific quota, irrespective, within limits, of the bulk of cellulose it is incidentally required to ingest. Monitoring of the passage of food in the mouth and buccal cavity does not seem to contribute to intake regulation, for if the oesophagus is exteriorized so that food swallowed does not enter the stomach, feeding causes only temporary satiety. Conversely, rats trained to feed themselves *directly* into the stomach by operating a syringe-driven intragastric tube, regulate their intake normally.

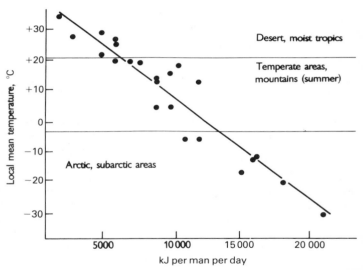

Fig. 5–3 Voluntary caloric intake of soldiers stationed in climates of different temperatures. (From JOHNSON, R. E. and KARK, R. M. (1947), *Science*, **105**, 378.)

(b) THERMAL SIGNALS Food intake is related to environmental temperature as can be seen from Fig. 5–3 which shows the voluntary food intake of soldiers stationed in a range of climatic conditions (recall the First Law of Thermodynamics and consider the reason for this). This observation, and the fact that heat production increases after a meal (the Specific Dynamic Action of food), led physiologists to speculate whether the satiety signal might be thermal: a slight increase in hypothalamic temperature consequent upon feeding might cause satiety and, furthermore, this mechanism might be reset by environmental temperature. This basic rather naïve concept that an animal eats to keep warm is now largely discounted, principally on the basis of one simple and elegant series of experiments. For, although normally there is indeed a slight rise in hypothalamic temperature during feeding as the hypothesis would require, if an animal is fed very cold food, it still regulates its intake accurately *despite* the fact that hypothalamic temperature now *falls* during feeding.

(c) CHEMICAL SIGNALS   As seen in Chapter 4, feeding results in large changes in the metabolic pools of energy substrates such as glucose, amino acids and FFA, together with complex changes in hormone balance. Do any of these changes influence feeding and satiety? The study of this problem has engendered considerable controversy and has probably posed more questions than it has answered. There are two major theories to be considered: the 'glucostat' theory and the 'lipostat' theory.

5.1.3   The 'glucostat' theory

This theory proposes that the satiety centre acts as a 'glucostat' and controls feeding by reference to the availability of glucose to it. Unlike the rest of the brain, the satiety centre requires insulin for glucose entry, thus the availability of glucose to the centre will depend upon both the blood glucose concentration and the amount of insulin in the circulation. As we saw in Fig. 4–3 the digestion and absorption of food would be expected to augment the availability of glucose to the satiety centre by increasing both the glucose and the insulin present in the blood. Evidence to support the 'glucostat' theory is widely available. Hypoglycaemia causes the sensation of hunger in man and promotes feeding behaviour in animals. Patients with diabetes mellitus, where insulin secretion is abolished, are hyperphagic: even though blood glucose is high, in the absence of insulin it cannot enter the satiety centre. However, further evidence suggests that the 'glucostat' theory cannot adequately explain all aspects of feeding regulation. Animals fed a diet of only protein and fat demonstrate little fluctuation in blood glucose and insulin secretion but nevertheless accurately regulate their food intake.

5.1.4   The 'lipostat' theory

This theory proposes that the guide-line an animal uses to regulate its energy intake is the condition of its fat stores. In conditions of fasting the fat reserves are being depleted and lipolysis is extensive, conversely, after a meal there is lipogenesis as fat absorbed from the diet is deposited in the fat stores. How the hypothalamus is able to monitor the condition of the fat stores and determine the relative extent of lipolysis and lipogenesis remains uncertain, although it has been suggested that specific, and as yet unidentified, factors may be released into the plasma: the lipogenetic factor would stimulate the satiety centre and the lipolytic factor, the feeding centre. In the latter connection, growth hormone has been suggested as the link since it is present in high concentrations during fasting, stimulates lipolysis and will, if injected, cause feeding.

The mechanisms that regulate food intake are still the subject of extensive research, particularly that directed towards the identity of the satiety signal. Fig. 5–4 summarizes the possible homeostatic mechanisms discussed above. When examining this figure bear in mind the possibility

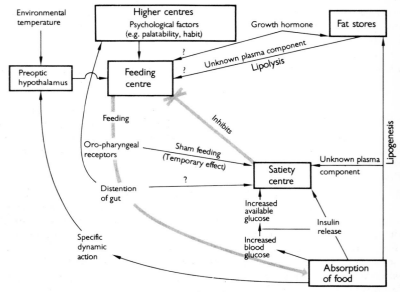

Fig. 5–4 Summary of possible pathways involved in the control of feeding (see text).

that the control of energy intake may well be a multifactorial phenomenon. Many factors, such as thermal signals, information from the gastrointestinal tract and chemical changes in the blood, may *contribute* toward overall control: there may be no single factor which can provide a complete explanation.

5.2 The regulation of deep body temperature

Mammals and birds are unique among living organisms in that they can regulate the temperature of the core of the body to within narrow limits in the face of wide fluctuations of environmental temperature. This ability to regulate deep body temperature devolves simply upon balancing the equation

$$\text{Heat production} = \text{Heat loss}$$

We will first consider the mechanisms available for the production and loss of heat and then consider the way in which balance between these factors is achieved.

5.2.1   Heat production

The great majority of chemical reactions in the body are exothermic, thus, even at rest, there is continuous production of heat: the 'basal metabolic rate' (BMR). There are a number of means by which this minimal heat production can be increased. These are summarized below:

- (i) Shivering
- (ii) Muscular exercise
- (iii) Calorigenic action of hormones: adrenaline causing a rapid short-lived increase in metabolic heat production, thyroxine causing a slower more prolonged effect
- (iv) The newborn of many species, including man, possess pads of specialized adipose tissue (brown fat) which can be activated by sympathetic nerves to produce heat through oxidation of FFA. This 'electric blanket' is also well developed in hibernating animals.

5.2.2   Heat loss

Heat is lost from the skin by radiation and convection and to a lesser extent by conduction. Further heat loss from the skin and the respiratory tract occurs by evaporation. Evaporative heat loss assumes crucial importance when environmental temperature exceeds skin temperature for, under these conditions, other physical means of losing heat actually serve to cause transfer of heat to the body. Evaporative heat loss from the skin occurs principally as a result of thermal sweating; activity of sweat glands brought about via their sympathetic nerves. Increased evaporative heat loss from the respiratory tract occurs during thermal panting: rapid shallow respiration associated with profuse salivation. Thermal panting is well developed in birds and fur-covered mammals such as cats and dogs, which do not possess sweat glands.

Heat loss by radiation, conduction and convection depends upon skin temperature, which in turn depends upon the blood flow through the skin and thus the transfer of heat from the core of the body to the surface. Blood flow through the skin is controlled by sympathetic vasoconstrictor nerves, variation in the discharge of which can cause both constriction and dilation of blood vessels (see Fig. 2–2).

Figure 5–5 illustrates the way in which blood flow through an appendage can be modified to promote either heat loss or heat

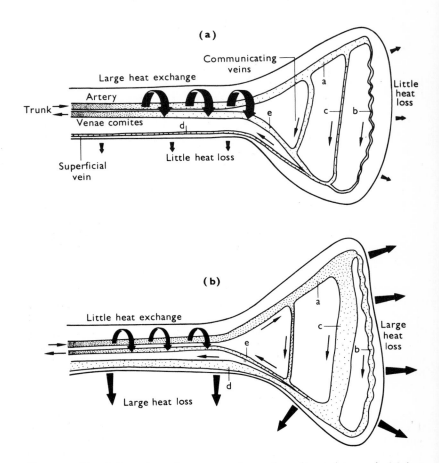

Fig. 5-5 Blood flow through an appendage when directed towards (a) *heat conservation*: a, Absolute flow through superficial tissues low. Thermal conductivity equivalent to cork. b, Flow within skin restricted to capillaries. c, Little flow in arterio-venous anastomoses (shunt pathways). d, Little flow in superficial veins. e, Cold blood returns from the extremities in the venae comites. It is warmed by counter-current heat exchange from the arterial blood; and (b) *heat loss*: a, Absolute flow through superficial tissues high. b, Some increase in capillary flow. c, Large blood flow through arterio-venous anastomoses. d, Most blood returns through superficial veins. e, Little blood returns through venae comites. Intensity of shading indicates temperature of blood, width of vessels indicates relative blood flow. (From HARDY, R. N., (1972). *Temperature and Animal Life*. Studies in Biology series, Edward Arnold, London.)

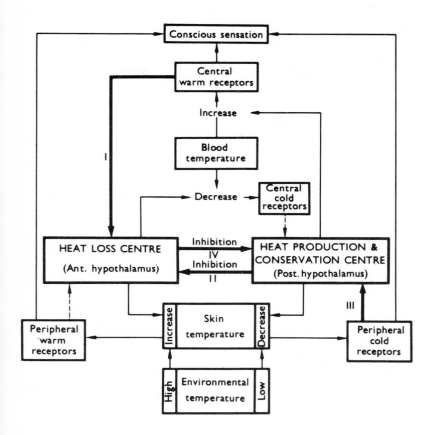

Fig. 5–6 Diagram to show the possible interrelationships between blood and environmental temperatures and the hypothalamic centres. Probable sources of conscious thermal sensation are also indicated. Important controlling pathways are indicated by dark lines. I: stimulation of anterior centre by hypothalamic warm receptors, II: inhibition of anterior centre by impulses from peripheral cold receptors acting via the posterior centre, III: stimulation of posterior centre by peripheral cold receptors and IV: inhibition of the posterior centre by central warm receptors acting via the anterior centre. Dotted lines indicate pathways of doubtful significance. (From HARDY, R. N. (1979), *Temperature and Animal Life*, 2nd edition, Studies in Biology series, Edward Arnold, London.)

conservation. Heat loss can also be reduced by the erection of hair or feathers covering the skin (piloerection). The hair is pulled into a more upright position by contraction of small slips of smooth muscle (piloerector muscles) which are stimulated by sympathetic nerves. In the erect position the hairs trap a thicker insulating layer of stagnant air around the body.

5.2.3 Central thermoregulatory control

The 'thermostat' responsible for modulating heat loss and heat production is located in the hypothalamus. Evidence from surgical lesions, electrical and thermal stimulation and recordings of the electrical activity of individual brain cells points to the existence of two 'centres'. The anterior part of the hypothalamus contains a group of cells designated the 'heat loss centre'. This centre promotes heat-dissipating responses such as sweating, panting and vasodilation in the skin and serves to inhibit the heat production and conservation centre in the posterior hypothalamus. This latter centre initiates heat-producing responses such as shivering and the release of thyroxine and adrenaline and serves to minimize heat loss by causing vasoconstriction in the skin and piloerection. Activity of this centre also inhibits the heat loss centre.

The two centres respond to changes in the temperature of the blood perfusing the hypothalamus and are also influenced by skin temperature. As a first approximation, the heat loss centre responds directly to an increase in hypothalamic temperature while the heat production and conservation centre responds largely as a result of activity in cold receptors in the skin. The overall arrangement of this sophisticated regulatory system is illustrated in Fig. 5–6.

5.3 Regulation of the volume and osmolality of extracellular fluid

The final section in this book is concerned with water and electrolyte balance within the mammalian body. There are two reasons for choosing to end with this topic: first because it is both complex *and* very important to an understanding of basic physiology, and secondly, because it provides a good example of how the simple theoretical concept of a control system can be investigated and thereby revealed to be much more complex in real life! You will appreciate the strategy of this approach if you compare Fig. 5–7b with Fig. 5–11 but do not be deterred, as we shall try to make the transition in easy stages.

5.3.1 Basic principles

In a normal subject the volume and total electrolyte concentration of the extracellular fluid (ECF) remain conspicuously stable in the face of a wide variation in external circumstances: indeed such stability is essential to normal function. Thus, if the total electrolyte concentration

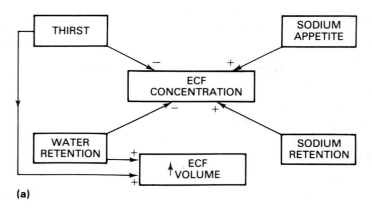

(a)

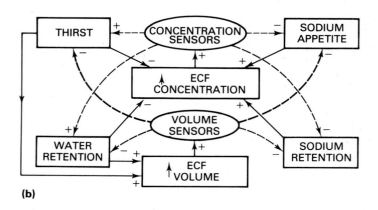

(b)

Fig. 5–7 (a) Diagram to demonstrate the interrelations between the intake and excretion of water and solute (sodium) and the volume and solute concentration of extracellular fluid (ECF). (b) As for (a), but with the addition of sensors measuring ECF volume and solute concentration and the way in which sensor information would modulate intake and loss to ensure homoeostasis. + = stimulation or increase. − = inhibition or reduction. Broken lines indicate pathways carrying sensory information: sense organs in oval boxes.

and therefore the osmolality of the ECF were to become disturbed, potentially damaging osmotic movements of water between cells and their environment would ensue, while gross changes in the volume of the ECF would impose grave problems for the cardiovascular system.

In essence, the regulation of the volume and composition of ECF is the result of a complex integration between neural factors controlling the intake of water and sodium (the principal cation) and endocrine factors controlling the excretion of these two substances. The basic layout of the control system can conveniently be visualized as in Fig. 5–7a. This figure demonstrates the interrelations between intake and loss of water; intake and loss of solute (sodium); and the volume and concentration of ECF. In order to maintain the stability of the volume and concentration of ECF there must be a means of monitoring each of these variables. Figure 5–7b contains such monitoring devices in the form of concentration sensors and volume sensors, stimulated respectively by an increase in the concentration or the volume of ECF. The dotted lines in Fig. 5–7b demonstrate the way in which information from the two types of sensors would need to influence the intake and loss of both sodium and water in order to ensure that the volume and electrolyte concentration of ECF is maintained within narrow limits. Careful examination of Figs 5–7 (a) and (b) should help to clarify the general principles underlying the control of water and electrolyte metabolism, but infortunately the *actual* physiological mechanisms used to bring about this control are much more complex and will require detailed analysis. In order to do this we shall first examine the control of water balance in isolation, then we will examine the control of sodium balance in isolation and then, finally, we will attempt to show the ways in which the two control mechanisms are integrated in order to achieve water and electrolyte homeostasis.

5.3.2   Regulation of water balance

Between 50 and 70% of body weight is water and of this about two thirds is contained within cells as intracellular fluid while the remaining third (some 20% body weight) is ECF which comprises plasma and tissue fluid (interstitial fluid). Normally the total body water remains relatively constant and its volume is regulated by ensuring the constancy of plasma volume, which of course provides an index of ECF volume. As was seen in Fig. 5–7a the ECF volume at any one time reflects the algebraic sum of water intake and water loss. It is therefore necessary to consider briefly the factors influencing water loss and water intake.

Control of water loss   Water is lost from the body by four routes; the kidney, the respiratory tract, the skin and the alimentary tract. Loss through the last three routes is *incidental* to other physiological functions (i.e. breathing, thermoregulation and digestion) and, as such, need not concern us in our consideration of the *control* of water balance. Excessive water loss through sweating, panting or gastro-intestinal malfunctions such as diarrhoea or vomiting will obviously temporarily disrupt water balance but they are not primary factors in its *regulation*.

The kidney is the organ controlling water loss and within the kidney the control is exerted *via* the effect of antidiuretic hormone (ADH), which promotes the reabsorption of water from the renal tubules. In the total absence of ADH virtually all the water entering the distal convoluted tubules passes into the urine and, since in man this represents some 20% of a glomerular filtrate of 130 ml. min^{-1}, enormous urinary water losses ensue (the production of an abnormally large volume of urine is termed a *diuresis*). Conversely, during extreme antidiuresis occasioned by maximal secretion of ADH, virtually all the water entering the distal tubule is re-absorbed along the tubule or from the collecting duct and consequently a very small volume of concentrated urine is produced and water loss from the kidney is minimized. Reference to our hypothetical control system in Fig. 5–7b will indicate that water retention (i.e. ADH secretion) should, in theory, be promoted by both an increase in ECF solute concentration (osmoregulation) or by a fall in ECF volume (volume regulation): evidence for the existence of such mechanisms will now be considered.

(*i*) *ADH and osmoregulation*   The classic work of Verney and his colleagues provided the first evidence that ADH secretion was stimulated by an increase in plasma osmolality. This stimulus would, of course, tend to cause water to pass from cells into the hypertonic ECF surrounding them and would thus tend to cause 'cellular dehydration'. Verney showed that infusions of hypertonic saline into the carotid arteries of conscious trained dogs caused a prompt inhibition of established water diuresis. The effect was similar to that produced by injection of posterior pituitary extract and could be evoked by hypertonic saline infusions which increased the osmolality of blood reaching the brain by as little as 2%.

Verney speculated that the osmolality of plasma was monitored by 'osmoreceptors' and, since intra-arterial hypertonic saline infusions were ineffective other than when directed toward the brain, concluded that the osmoreceptors were within the brain. Recent work, in which minute volumes of hypertonic solutions were injected directly into the brain, has shown that the osmoreceptors are found lateral to the supraoptic nucleus (SON) and thus are distinct from the actual cells that produce ADH itself (Fig. 5–8). Even more recently, some doubt has been cast on the classical concept of osmoreceptors in the supraoptic region and it has been suggested that the principal changes in ADH secretion in response to fluctuation in plasma solute are mediated via a specific Na^+ sensor on the surface of the brain adjacent to the third ventricle, which responds to changes in the Na^+ concentration in the cerebrospinal fluid (CSF). The situation is still unclear so both possibilities are shown in Fig. 5–8.

(*ii*) *ADH and volume regulation*   A reduction of 10% or more in blood

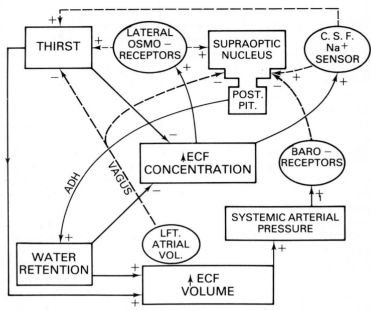

Fig. 5–8 Diagram to illustrate the pathways by which the volume and solute concentration of extracellular fluid (ECF) may regulate water intake and the secretion of antidiuretic hormone (ADH) from the posterior pituitary gland. CSF, cerebro spinal fluid. + = stimulation or increase; − = inhibition or decrease. Broken lines indicate nerves or nerve tracts carrying sensory information: sense organs in oval boxes.

volume (hypovolemia or 'extracellular dehydration') is an effective stimulus to ADH secretion. The receptors responsible for initiating this neuroendocrine reflex are the stretch receptors in the left atrium and, to a lesser degree, the arterial barorecptors (Fig. 3–1). The left atrial receptors have been investigated by inserting balloons into this heart chamber and demonstrating that distension results in a decrease in plasma ADH concentration and consequently an increased flow of urine. Conversely, if blood volume is decreased as, for example, after haemorrhage, the atrial stretch receptors are less active thus ADH secretion and water retention will result and help to restore plasma volume.

Impulses from the left atrial receptors travel to the CNS via the vagus nerves and ultimately cause a decrease in activity in the SON (Fig. 5–8). Increase in plasma volume thus inhibits ADH secretion by increasing activity in the vagal afferents, while decrease in plasma volume causes increased ADH secretion by decreasing the *inhibitory* effect of vagal impulses from left atrial receptors. Baroreceptors on the arterial side of the circulation have also been shown to influence ADH secretion, but it appears that this system is much less sensitive than the left atrial receptors and probably only comes into action under very extreme conditions where the changes in blood volume are sufficiently great to be reflected in pronounced changes in blood pressure.

(*iii*) *Interactions between osmoregulation and volume regulation in the control of ADH* The osmolality of plasma in the hypothalamus exerts the sensitive primary control over ADH secretion, but there is a parallel but less sensitive control pathway determined by plasma volume as monitored by the stretch receptors in the left atrium. In circumstances such as dehydration, both these control pathways, responding respectively to the increased osmolality and the decreased blood volume, reinforce each other in stimulating ADH secretion. In other circumstances, however, such as severe sodium depletion, the osmoreceptors are inhibited as the plasma osmolality is low. However, if ADH secretion were inhibited in order to restore normal osmolality, the resultant water loss would tend to cause a dangerous fall in plasma volume and thus the volume receptor mechanisms would come into action to stimulate ADH. In such circumstances the volume is maintained in preference to the osmolality since the circulatory failure which would result from a severe fall in plasma volume would be more deleterious than the low plasma sodium concentration. In practice, the situation is much more complex and is interrelated with the control of sodium excretion by aldosterone (see p. oo).

Control of water intake As will be seen from Fig. 5–8 the stimuli to thirst are in most respects comparable with those for ADH. Thus, increase in ECF solute concentration promotes thirst via osmoreceptors in the lateral preoptic hypothalamus (probably anatomically separate from those influencing ADH) and also probably via the Na^+ sensor in the CSF, while changes in ECF volume, as indicated by volume receptors in the left atrium, also modulate drinking. The peptide angiotensin II (see p. 60) is also known to stimulate both drinking (dipsogenic effect) and the secretion of ADH (Fig. 5–11). Additionally, in many species, temporary satiety results from the monitoring of fluid intake during drinking: this gastropharyngeal metering is remarkably accurate so that animals drink only sufficient fluid to restore their fluid deficit.

5.3.3   *Regulation of sodium balance*

Sodium ions account for more than 90% of the total osmotically effective cation in ECF. Moreover, whenever cations are absorbed into the ECF from the gut lumen or the renal tubules, the associated movement of electrical charge results in the transfer of an equivalent amount of anions. It follows from these two statements that at least 90% of the total osmolality of ECF is directly or indirectly determined by sodium ion concentration [Na$^+$]. Since we have just considered the exquisitely sensitive way in which ADH secretion and drinking are related to ECF osmolality or [Na$^+$] it follows that changes in ECF sodium content will exert far-reaching effects on ECF volume. Consequently, the control of sodium balance is a major factor in electrolyte and volumetric homeo-stasis and once again it resolves itself into a question of the balance between intake and loss (*q.v.* Fig. 5–7).

Control of sodium loss   Sodium is lost from the body in sweat, faeces and urine and of these routes only loss through the kidney can be regulated as directed by the requirements of sodium balance. Renal handling of sodium and its control is one of the most controversial areas in physiology, the technical problems of the study of which are only matched by the fundamental clinical importance of understanding the mechanisms involved.

We shall restrict our discussion to consideration of the well-established role of the adrenal mineralocorticoid hormone aldosterone in pro-moting sodium retention by the kidney because available evidence indicates that this action of aldosterone, although not the only factor in sodium excretion, is probably the most significant.

(*i*) *Control of aldosterone secretion*   The most important factor in stimulating the secretion of aldosterone by the adrenal cortex is the octapeptide angiotensin II, which is derived ultimately from a plasma protein angiotensinogen produced by the liver. The rate-limiting step in angiotensin II production (and thus aldosterone secretion) is the presence of the enzyme renin which is secreted from the juxta-glomerular apparatus (JGA) in the kidney (Fig. 5–9). This so-called renin-angiotensin system has been the subject of enormous research effort, both because of its fundamental role in aldosterone secretion and hence salt and water balance, but also because its components (renin and angiotensin II) are among the most powerful vasoconstrictor substances known and thus may play a part in certain forms of cardio-vascular disorder associated with high blood pressure (hypertension).

In essence, the control of aldosterone secretion and sodium excretion resolves into the question of what regulates the release of renin from the JGA. A glance at Fig. 5–10 will reveal that the answer appears far from simple and indeed it should be stressed that this diagram is itself a gross oversimplification. It is well established that the following circumstances provoke renin secretion:

(*a*) A decrease in blood pressure within the kidney (e.g. following a decrease in systemic arterial pressure or narrowing of the renal artery).

(*b*) Decrease in ECF volume (e.g. after haemorrhage).

(*c*) Sodium depletion (low sodium intake or the use of diuretics causing sodium loss).

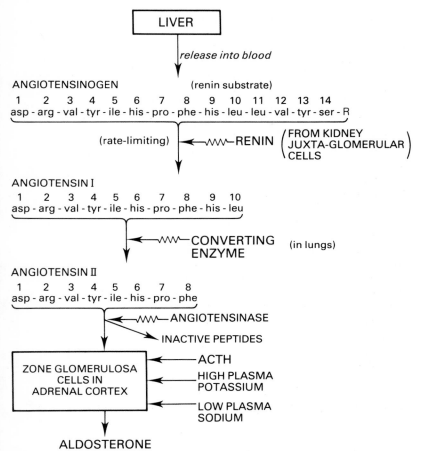

Fig. 5–9 The renin-angiotensin system and the control of aldosterone secretion. (Modified from HARDY, R. N. (1982). *Endocrine Physiology.* Physiological Principles in Medicine Series, Edward Arnold, London.)

(*d*) Drugs which mimic the action of the sympathetic nerves to the kidney.

In order to explain how these factors act to stimulate renin release by the JGA three parallel effector control pathways have been suggested, they are (numbered as in Fig. 5–10):

① Fall in renal arterial pressure.

② Decrease in the 'filtered sodium load': this is the rate at which sodium arrives at the tubules and is the product of the plasma sodium concentration [Na^+] and the rate of plasma filtration (glomerular filtration rate: GFR).

③ The action of the renal sympathetic nerves.

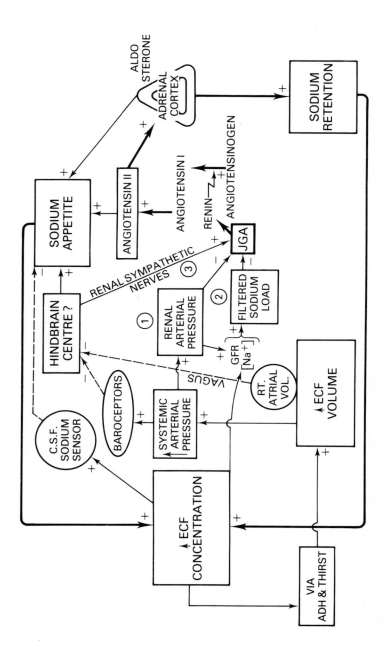

Fig. 5-10. Diagram to illustrate pathways involved in the regulation of sodium balance. GFR, glomerular filtration rate; CSF, cerebrospinal fluid; JGA, justa-glomerular apparatus. + = stimulation or increase; − = inhibition or decrease. Broken lines indicate nerves or nerve tracts carrying sensory information: sense organs in circular or oval boxes.

Look carefully now at Fig. 5–10 and you should be able to trace how the stimuli to renin release tabulated above *a–d* will each activate one or more of the effector paths ①, ② and ③. It should now also become clear why the study of the renin-angiotensin system is so different – particularly when it is noted that other factors listed briefly below *also* effect aldosterone secretion *and*, as we shall see in the final section, because we have in addition to integrate the control of aldosterone with that of ADH and thirst.

Other factors which may effect aldosterone secretion include; ACTH (Table 1) when present in high concentration during stress (N.B. fall in blood pressure or haemorrhage is a stress); increase in plasma potassium, which stimulates aldosterone by direct action on the adrenal cortex (this makes sense because aldosterone causes increased sodium reabsorption by the kidney *in exchange* for potassium – hence aldosterone promotes potassium excretion).

Control of sodium intake   In human subjects on a normal diet, sodium intake is in excess of requirements and thus the concept of a specific *sodium appetite* has little relevance. However, under experimental conditions, deliberate restriction of sodium intake by salt-free diets does allow study of sodium appetite. Work on experimental animals has shown that sodium appetite is stimulated by aldosterone and possibly angiotensin II which are both substances present in the circulation during salt deficiency. Conversely, sodium appetite is depressed if ECF sodium concentration or volume are high, probably by the pathways shown in Fig. 5–10.

5.3.4 Recapitulation

We have seen how the water balance of the body is maintained by alterations in water intake and water loss in response to changes in ECF volume and solute concentration (Fig. 5–8) and also how the latter factors influence sodium intake and loss (Fig. 5–10). It remains to try to integrate the essential features into a composite whole in order to be able to visualize the overall strategy for water and sodium homeostasis. Figure 5–11 is an attempt to achieve this objective: it is a complex diagram but should not be difficult to find your way around if you have understood the basis of Figs 5–7 and 5–10. Finally, return to Fig. 5–7 in order to compare how the body has achieved in practice a control system with the characteristics we specified in our basic model.

5.4   The role of the hypothalamus in homeostasis

All complex homeostatic systems involve the hypothalamus and when one examines the input to and output from this area of the brain the reason for its implication in homeostasis becomes immediately apparent.

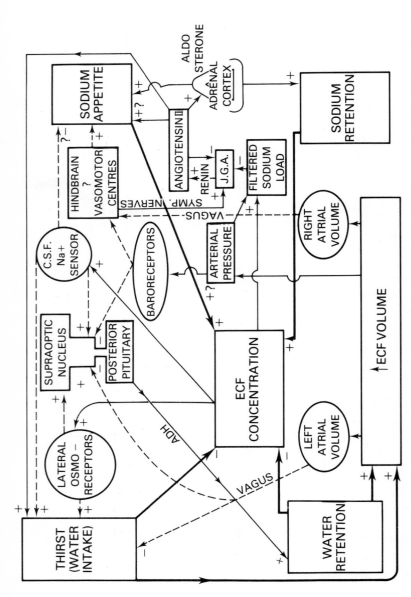

Fig. 5-11 Composite diagram to illustrate the overall integration of factors involved in water and sodium homeostasis. CSF, cerebrospinal fluid; JGA, juxta-glomerular apparatus. + = stimulation or increase; − = inhibition or decrease. Broken lines indicate nerves or nerve tracts carrying sensory information; sense organ in circular or oval boxes.

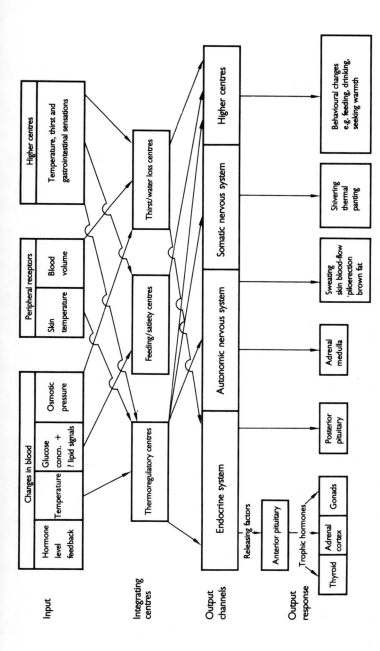

Fig. 5-12 Overall view of the position of the hypothalamus in relation to sensory information (input) and effector pathways (output) with specific reference to thermoregulation, energy balance and fluid regulation.

It is unique in the access it has to the effector systems of the body and also in terms of the sensory information it monitors. In addition, it contains a number of control 'centres' which are primarily responsible for integrating sensory information and producing the appropriate executive responses. Figure 5–12 attempts to illustrate the central role of the hypothalamus in homeostasis as exemplified by thermoregulation and the control of food and water balance.

6 Conclusion

Walter Cannon who coined the term homeostasis in 1929 would be gratified and amazed by the progress made during the last 45 years in the analysis of the homeostatic systems with which he was so familiar. In large measure this progress, as with the advance of all science, reflects the advances in scientific technique: so often in research the question to ask is obvious, while the means to achieve an answer are seemingly impossible. Impossible that is for a time, for in many cases sooner or later someone develops the necessary technique. It is doubtful whether Cannon could have conceived that it would ever be possible to record the activity of individual cells in the brain. However, with the development of microelectrodes, stereotactic surgery and sensitive electronic recording apparatus this is now a routine technique and is even sometimes used in undergraduate practical classes. He would certainly have speculated that it would eventually be possible to measure accurately and specifically the hormone concentrations in blood plasma, but could he have foreseen the sophistication of modern radioimmunoassay techniques? With these it is possible, for example, to measure in samples of 10 μl of blood the concentration of a hormone present only in pg quantities (1 picogram $= 10^{-12}$g) in 100 ml blood.

Throughout this book you will have been conscious of how little is known of the physiology of invertebrates in relation to the wealth of information about mammalian function. In the future this imbalance is bound to persist, because the much greater potential relevance of mammalian physiology to medicine means that correspondingly greater research resources will be devoted to it. Nevertheless, it is to be hoped that, as in the past, the application to the invertebrates of techniques developed principally for mammalian physiology will ensure that our understanding of invertebrate function will continue to improve.

Further Reading

BARRINGTON, E. J. W. (1979). *Invertebrate Structure and Function.* (2nd edn.). Van Nostrand Rheinhold, Wokingham.

BARRINGTON, E. J. W. (1968). *The Chemical Basis of Physiological Regulation.* Scott, Foresman & Co., Illinois.

CANNON, W. B. (1963). *The Wisdom of the Body.* Norton, New York.

CHAPMAN, R. F. (1975). *Insects: Structure and Function.* Hodder, London.

HARDY, R. N. (1972). *Temperature and Animal Life.* Studies in Biology no. 35. Edward Arnold, London.

LANGLEY, L. L. (1973). *Homeostasis, Origins of the Concept.* Dowden, Hutchinson & Ross, Inc. Stroudsburg.

MOUNTCASTLE, V. B. (ed.) (1980). *Medical Physiology* (14th edn.). C.V. Mosby, St. Louis.

PROSSER, C. L. (1973). *Comparative Animal Physiology* (3rd edn.). W.B. Saunders Company, Philadelphia and London.

TOATES, F. M. (1975). *Control Theory in Biology and Experimental Psychology.* Hutchinson, London.

WIENER, N. (1961). *Cybernetics* (2nd edn.). MIT Press, Cambridge, Mass.

WIGGLESWORTH, V. B. (1974). *Insect Physiology* (7th edn.). Chapman and Hall, London. ˙

Index